Teachers' Pack on Experiments in Materials Science

Dr Claire Davis
University of Birmingham

VIDEOTAPE
Please note that the videotape which was previously supplied with this booklet is no longer available.

Routledge
Taylor & Francis Group

LONDON AND NEW YORK

FOR THE INSTITUTE OF MATERIALS

Acknowledgements

I would like to thank the following teachers who gave me valuable feedback on this pack:

Steve Clark	*South Bromsgrove High School, Bromsgrove*
Andrew Davies	*St Boniface's College, Plymouth*
Keith Moseley	*Monmouth School, Monmouth*
Krystyna Smith	*Witchford Village College, Nr. Ely*
Richard Wall	*Willenhall Comprehensive, Willenhall*
Tony Woods	*St Peters Collegiate School, Wolverhampton*

I would also like to thank the following for their valuable comments:

David Hartley	*Subject Officer for GCE Physics, EDEXCEL Foundation*
Susan Oldcorn	*Examiner for Materials Option, EDEXCEL Foundation*
Miranda Stephenson	*Chemical Industry Education Centre, University of York*

The assistance and ideas of Dr Paul Withey are gratefully acknowledged as are the ideas from Dr Martin Strangwood, Dr Nigel Mills, Dr Robin Grimes, Dr Ian Hutchings and Dr James Marrow. The financial assistance of the School of Metallurgy and Materials, The Worshipful Company of the Armourers and Brasiers, British Steel and The Institute of Materials is gratefully acknowledged.

Contents

Introduction

This pack is designed as an aid to teachers having to teach materials aspects in an 'A' level physics, chemistry, design and technology or GNVQ general science course.

The aim is to provide suggestions for experiments to accompany course material. Information in the pack may be photo-copied and the experiments integrated into the course as appropriate. There are also a few data handling exercises given with data sets for homework or class activities where carrying out the experiment is not practical. All the experiments can be carried out using readily available equipment/materials and generally require little preparation.

The pack is laid out in an easy to follow format. There are four chapters tackling different general topics -

> Materials and Structure
> Mechanical Properties
> Processing
> Materials Selection

Within each chapter there are sections covering different material types, test techniques, processes etc. To help in the planning of lessons etc. **all experiments are written in bold** and *all examples are given in italics* . In addition there are photographs of microstructures and load-displacement curves from tensile tests to be used in association with some of the experiments.

I hope that this pack will prove to be useful to science teachers both for formal teaching and during club activities. Metallurgy and Materials Science is an extremely interesting and diverse subject with excellent career opportunities. All of the modern technology around us is somehow linked to engineering and materials science - I hope that this pack will open the eyes of some of your students to the possibilities of the subject.

Please feel free to contact me with any additional experiments that you feel should be included or with any questions you might have on where to find extra information.

School of Metallurgy and Materials
University of Birmingham
Edgbaston
Birmingham
B15 2TT
Tel. 0121 414 5174
Fax. 0121 414 5232
E-mail c.l.davis@bham.ac.uk

Dr Claire Davis

MATERIALS AND STRUCTURE
List of Experiments

CHAPTER I
MATERIALS AND STRUCTURE

1.1 Materials Types

For the first 92 elements in the Periodic Table there are three classifications which can be used: metals, non metals and inert gases. But from all of these over 300 000 different materials can be produced. These materials can be classified into four broad categories: metals, polymers, ceramics and glasses, and composites. In terms of the engineering materials they can be considered as in Figure 1.1.

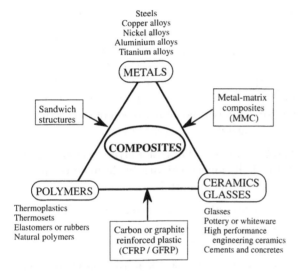

Figure 1.1 The classification of engineering materials.

In this section we will mention these classification types and some general examples.

1.1.1 Metals

A metal can be thought of as a solid or liquid held together by metallic bonding and generally shows characteristic properties of high reflectivity, high electrical and thermal conductivity and relatively high density compared with non-metals.

Metals can be thought of as the pure metal element such as gold, iron, aluminium etc. or as an alloy where an alloy is a mixture of elements which has metallic properties, for example steel, brass, bronze etc.

Typical examples of metals that are met in everyday life are:

copper– *high purity copper is used as an electrical conductor in domestic wiring*

aluminium – *commercially pure in aluminium foil, milk bottle tops*

gold, platinum and silver – *in jewellery*

mercury – *in thermometers*

chromium – *chrome plated taps, car bumpers*

nickel – *plated on bra strap fasteners, watch straps, cheap jewellery*

zinc – *zinc plating - i.e. galvanising e.g. lamp posts*

tungsten – *lamp filaments, dart tips*

lead – *roofing sheet, e.g. on churches*

magnesium - *distress flares, fireworks*

Typical examples of alloys that are met in everyday life are:

steel – *there are thousands of different steels used commercially e.g. bridges, car bodies, stainless steel sinks, cutlery, building materials, coinage etc.*

aluminium alloys – *again hundreds - e.g. aeroplane skins, drinks cans,*

titanium alloys – *fan blades in aero-engines, body implants (pace makers, hip joints etc.)*

nickel alloys – *turbine blade in aero-engine, leading edge of helicopter rotors, shoe moulds*

copper alloys – *brasses, bronzes - coinage, screws, door knobs*

tin–lead alloy – *solder*

gold alloys – *alloyed with silver or platinum in jewellery, electrical connections to silicon chips e.g. in computers, calculators etc.*

1.1.2 Polymers

A polymer can be thought of as a solid or liquid consisting of large molecules, comprised of repeating units (monomers), which generally consist of a carbon chain. Polymers can be divided into thermoplastics and thermosets (see glossary).

Typical examples of polymers that are met in everyday life are:

PVC *(polyvinyl chloride)* – *window frames, pipes*

PE *(polyethylene)* – *washing up bowls*

Nylon – *ropes, some camping equipment, tights*

Polyester – *shirts*

PP *(polypropylene)* – *parcel wrapping*

PET *(polyethylene terephthalate)* – *lemonade bottles*

PC *(polycarbonate)* – *CDs*

PS *(polystyrene)* – *CD cases, plastic rulers, set squares etc. expanded PS used in bike helmets, packing material*

PTFE *(polytetrafluoroethylene)* – *non-stick coating on frying pans etc.*

Rubbers – *wellington boots, o-rings, erasers, shoe soles, condoms*

UHU glue – *example of a liquid polymer*

1.1.3 Ceramics and glasses

A ceramic can be thought of as a crystalline solid produced by the action of heat on a single or a number of crystalline inorganic non-metallic materials. A glass has an amorphous structure and is often based on silica, it is therefore not a true ceramic although shares many similar properties.

Typical examples of ceramics that are met in everyday life are:

Alumina (Al_2O_3) – *spark plugs, thread guides in textile factories*

Sapphire (*Co doped* Al_2O_3) - *lasers*

Tungsten carbide - *drill tips*

Cement – *construction*

Silicates – *electrical power insulators*

Diamond – *jewellery, tool tips*

Glass ceramics – *missile nose cones, microwave components, domestic cooker tops*

Silica glass – *telescope mirrors, windows, space shuttle tiles (see Video)*

Borosilicate glass (*Pyrex*) – *heat resistant cookware*

1.1.4 Composites

A composite can be thought of as a material which is made of two or more different types of materials in an intimate mixture.

Typical examples of composites are given below:

MMCs (*metal matrix composites*) – *metal matrix with particulate or fibre reinforcement, e.g. Al-SiC particulate composites used in automotive and aerospace applications, bicycle frames, brake callipers. Titanium composite in compressor discs in aero-engines.*

PMCs (*polymer matrix composites*) – *carbon or glass short fibre epoxy composite in sports equipment, for example squash racquets, fishing rods. Formula 1 car body. Aircraft body panels. Flexible magnets - magnetic particles in a polymer matrix.*

CMCs (*ceramic matrix composites*) – *concrete - composite of cement and sand/gravel -construction.*

EXPERIMENT 1.1: MATERIAL TYPES

Put together your own 'bag of bits', i.e. a collection of common materials/products that you can bring out and ask students to identify the materials/components and why they think they were selected for that application. This can be used as an interesting opportunity for discussion on the subject of materials, especially aspects such as selection, properties, recycling, cost etc.
Some examples are given below:

copper wires in a plug – **used because of their high conductivity**

steel waste paper bins – **cheap, formable and fire resistant etc., note that you may have plastic (polypropylene or high density polyethylene) or wood bins, the reasons are cost, availability, formability and aesthetics.**

PVC window frames or polystyrene rulers – **cheap and easily formed by injection moulding.**

Silica glass for windows – **as transparent + more scratch resistant than plastics etc.**

Silica glass or polycarbonate for spectacles – **can find examples using either – both transparent. Polycarbonate is less dense therefore have lighter spectacles but less scratch resistant so needs coating but also less likely to fracture hence use for safety glasses.**

Stainless steel scissors – **use very hard (martensitic - see experiment 3.5) steel to give a good cutting edge, + stainless to avoid rusting. Note: often have plastic handles (could be nylon, PP, HDPE).**

Clay floor tile – **kaolin-based clay chosen for cost, wear resistance, aethestics.**

Alumina spark plugs – **low electrical conductivity, hard.**

Tennis racket – **modern rackets will be made from composite materials such as carbon fibre reinforced plastic (CFRP) or kevlar. Could use rackets as an example of changing materials use over last few years in sports from wood and aluminium – used for low density, high stiffness, high strength. The strings have also changed from cat gut to nylon.**

Flexible magnets – **low cost. Polymer matrix is used for its flexibility and as a binder for the magnetic particles. i.e. a composite material.**

Drinks cans – **aluminium or steel (check with a magnet), also have a number of plastics containers, even cardboard for milk.**

Other examples can be taken from the examples given in sections 1.1.1 to 1.1.4. You can use fabricated products breaking them down into their many components. For example a bicycle pump, small electric motor, plug etc. You could initiate links with local industry to provide samples of products and manufacturing/materials details.

From this it is possible to pick out key selection criteria that are often required such as cost (of the material and for processing to the final product shape), strength, stiffness, formability, density, electrical/thermal conductivity, fire resistance, aesthetics and others. This could lead onto discussion of these aspects. The data sheet on mechanical properties and cost per kg. may be useful with this exercise. (Note for a direct replacement of, for example, steel by a plastic the cost of the material will be less because less weight will be required). You will often find that a component could be made out of a number of materials, and indeed may be, which can be discussed. For example chairs may be steel, wood, plastic or cast iron (especially garden chairs). It is then possible to explore the reasons.

Composite materials are less commonly seen as they are more expensive to produce, you generally find them where an increased performance justifies the increased cost such as the increased strength and stiffness advantage used in sports equipment. There are many examples where composite materials are chosen for their increased strength or stiffness advantage (from the reinforcement) combined with suitable toughness (from the matrix). It is more difficult to find examples of other reasons for their use (flexible magnets).

NOTE: care is needed to identify composite materials rather than composite structures, see also Experiment 1.11 (in Section 1.3.4) where we look at the structure of composite materials and the difference between composite materials and composite structures. It should be noted that this distinction is subjective as you could consider composite structures as composite materials if you wished.

1.1.5 Comparison between material types:

The different material types can be compared using several different techniques. The main mechanical methods such as hardness, strength, toughness, stiffness are described in Chapter 2. There are a number of physical methods that can be used, some of these are given below:

EXPERIMENT 1.2   COMPARISON OF MATERIAL TYPES – USING PHYSICAL PROPERTIES TO COMPARE THE MATERIAL TYPES

1.   Conduction of heat: You need rods of the same diameter of a selection of materials, for example copper, steel, ceramic such as alumina or silica (glass), wood, polyethylene etc. Take a block of metal approx. 60 mm x 60 mm x 30 mm - steel is best but aluminium would do. Drill three holes, about 20 mm into the block to fit the rods snugly (it is best to compare three materials at a time for increased accuracy - from uniformity of heating). Attach beads of wax at 10 mm intervals along each rod. Heat the metal block evenly and slowly with a Bunsen burner. Note the time at which the consecutive beads of wax melt on each rod. (E.g. the copper is considerably faster than the steel which should be considerably faster then the glass) The result can be displayed graphically.

2.   Conduction of electricity: You can use the same materials as in 1. to show which materials conduct electricity using a simple electrical circuit. As a general rule metals conduct whereas ceramics and polymers do not - care is needed as some metals do not conduct very well and some, such as aluminium, are covered in a thin insulating oxide (e.g. Al$_2$O$_3$ - a ceramic).

3.   Magnetism: This is less to tell material types apart but more an interesting side topic - the only metals that are magnetic are iron, cobalt and nickel. This means that steels should be magnetic ... this is true for the majority of steels (any mild steel which is used in numerous items such as bicycles, screws, keys, paper clips etc.). However there is one type of steel that is non-magnetic; austenitic stainless steels (note, however that some stainless steels are magnetic!). This is due to the

difference in crystal structure between austenitic stainless steel (face centred cubic) and mild steel (body centred cubic) - see section 1.2.1 for structure types. This makes for some interesting observations when trying to determine metal types - e.g. austenitic stainless steel is used in domestic appliances - sink tops, cutlery, parts in dishwashers, washing machines.

4.   An interesting experiment can be carried out utilising the different thermal expansivity of steel and brass. Take strips, measuring approx. 0.5 mm x 20 mm x 200 mm, of brass (copper will also work) and mild steel (i.e. not stainless). Attach these strips face to face using rivets or a nut and bolt at each end. Heat the strip with a Bunsen burner and watch it bend. The length of the brass expands much faster and more than that of the steel thereby pushing the strip over and curling up the bi-metallic strip.

1.2   Structure Types

The materials mentioned above have different atomic structures due to differences in bonding and/or arrangement of the constitutive components. For example a polymer can exhibit amorphous, or semi-crystalline structures.

A definition of a crystalline solid is a solid in which the atoms or molecules are arranged in a regular manner, the values of certain physical properties depending on the direction in which they are measured. A definition of amorphous is a structure without the periodic, ordered structure of crystalline solids.

Examples of crystalline materials seen in everyday life are:
 sugar, salt etc.
and examples of amorphous materials are:
 glass, rubber – *elastic bands, erasers etc.*

In this section we will consider the structures commonly observed in the different material types. We will be talking about structure on the atomic level here, in the next section we will deal with the microstructure of materials where the microstructure refers to the structure on a microscopic scale (about 1–1000μm in size).

1.2.1 Metals

Metals can be amorphous or crystalline but in all cases the bonding between atoms is metallic bonding. Metallic bonding is non-directional which allows atoms to pack closely together.

Amorphous metals are often referred to as metallic glasses and can be formed when a metal is cooled extremely quickly (quenched) from its molten state. Metallic glasses are by far and away the less common form of metals and are used as magnetic materials for example in video and audio tape heads.

Metals with a crystalline structure have the component atoms arranged in regular manner - for example aluminium has, at room temperature, what is called a close packed arrangement. This is where atoms, if assumed to be hard spheres, are packed as close together as possible (74% packing efficiency).

EXPERIMENT 1.3: CRYSTALLINE STRUCTURES

In two dimensions this can be represented by using some 1p coins and building up a regular array. A little more involved would be to use expanded polystyrene spheres or a bubble model (this can also be used to demonstrate dislocations - see section 1.4). The concept of close packed directions, directions in which the atoms are touching, can be introduced, Figure 1.2.

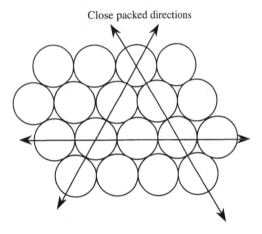

Figure 1.2 Close packed layer (also called a close packed plane) showing close packed directions.

In three dimensions close packed arrays can be of two types – face centred close packing or hexagonal close packed. These can be developed through different stacking arrangements:

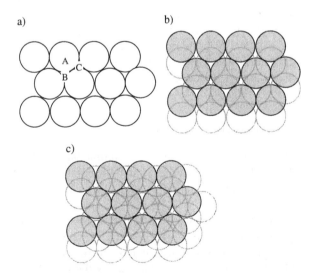

Figure 1.3 (a) Close packed layer showing positions for stacking, (b) hexagonal close packing (i.e. stacking sequence ABABA), (c) face centred close packing (i.e. stacking sequence ABCABCA)

Considering the first layer of the stack, Figure 1.3, then if the second layer is stacked with position A above position B and the third layer with position A above position A etc. then the arrangement is hexagonal close packing (h.c.p.). If, however, the third layer is stacked with position A over position C then the sequence of 3 repeated you have face centred close packing (f.c.c.).

Metals do not have to adopt a close packing arrangement, another common structure is that of body centred packing. Here the first layer consists of a regular array of atoms that are not touching. Another layer sits on top of these, as shown by the shaded circles in figure 1.4, with a further layer directly above the first one. The close packing arrangement (f.c.c. or h.c.p.) has a packing density, i.e. space filling efficiency, of 74% whereas the body centred cubic (b.c.c.) has a packing density of 68%.

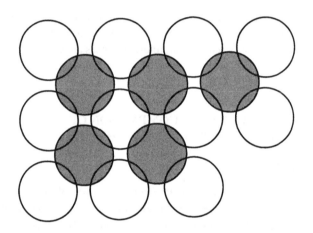

Figure 1.4 Schematic diagram of body centred stacking.

A single metal may adopt a number of different structures depending on temperature and pressure. For example iron can be either face centred close packed or body centred - it is body centred at room temperature and pressure but if the temperature is raised above 923°C it becomes face centred close packed.

These regular arrays can be observed using electron microscopy techniques – either high resolution electron microscopy (HREM) which reveals the atomic array (see Figure 1.21) or by diffraction patterns, Figure 1.5, which can be interpreted to give the structure.

Alloys can have structures that exhibit regular crystalline arrangements where the constituent atoms can occupy different positions within the lattice. *For example a simple plain carbon steel is an alloy of iron and carbon where the carbon atoms sit in the holes inbetween the face centred close packed or body centred iron structure.*

EXPERIMENT 1.4 CRYSTALLINE STRUCTURES – ALLOYS, DIFFERENT SIZE CONSTITUENTS

If you take a large glass beaker of ping pong balls and put in a handful of small spheres e.g. dried peas then shake it a little it can be seen that the peas (if sufficiently small) will sit in the gaps between the polystyrene spheres. If the peas are a little too large for the holes there is a slight distortion of the lattice – this is also seen for some alloys.

Another example is brass which is an alloy of copper and zinc. Both these atoms are of similar size and the structure of brass is body centred, in this case looking at Figure 1.4 then the Cu atoms would be the white spheres and the Zn atoms the shaded spheres. In some alloys the relative positions of the different elements is fixed whereas in others they are random.

a)

b)

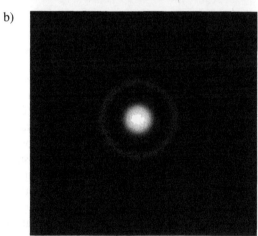

Figure 1.5 Electron diffraction pattern of (a) face centred structure showing the regular pattern indicating a crystal structure
(Note: the bright spots are diffraction spots, not atomic positions) and (b) an amorphous material (amorphous silica) showing rings with little/no regular pattern.

EXPERIMENT 1.5 CRYSTALLINE STRUCTURES – ALLOYS, SAME SIZE CONSTITUENT

Take a beaker and place in it two sets of the same size ping pong balls, one set of a different colour. If they are mixed together then the arrangements are the same only the relative positions of the 'elements' are different.

1.2.2 Polymers

Polymers can be fully amorphous or semi-crystalline. An amorphous polymer is one where the polymer chains cannot crystallise, i.e. form a regular, periodic arrangement. This may be due to the polymer being cooled rapidly to avoid crystallisation or because the irregular nature of the polymer chains (large side groups being arranged at random around the main carbon chain). In a semi-crystalline polymer the chains are partially aligned. Some polymers may have both amorphous and crystalline regions, figure 1.6.

Examples:

• *polystyrene (PS) is amorphous*
• *polyethylene (PE) is partially crystalline - HDPE (high density PE) can be ≈ 80% crystalline*

The bonding in polymers is covalent between the atoms within the polymer chains and there is Van der Waals bonding between the chains.

EXPERIMENT 1.6   AMORPHOUS AND CRYSTALLINE STRUCTURES IN POLYMERS

Pour ('cast') freshly cooked spaghetti into a glass beaker and leave to harden. (A fine spaghetti such as vermicelli gives a better result). It is often possible to see small areas where the spaghetti 'crystallises' with aligned lengths. It is better to wait a few days until the spaghetti is completely dried and then slice through the spaghetti as small 'crystallised' areas are more easy to see. You can calculate/estimate the amount of crystallised area - maybe compare the different diameter spaghettis.

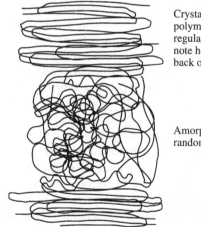

Crystalline region - polymer chains in a regular array, note how they fold back on themselves

Amorphous region - random chain coiling

Figure 1.6 (a) Schematic diagram showing crystalline and amorphous regions of a polymer.

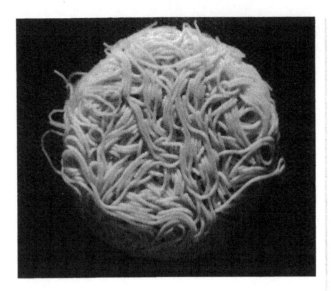

Figure 1.6 (b) Photograph showing crystallised and amorphous regions in spaghetti.

We have seen that polymers can contain amorphous and crystalline regions. It is also possible to align the polymeric chains in an amorphous region thereby making it crystalline. Chapter 3 section 3.2 discusses how the aligned chains present in samples which have been injection moulded (for example cheap polystyrene - 'plastic' rulers and protractors) can be viewed using polarised light. Experiment 1.7 illustrates how polymeric chains can be aligned using stretching or 'drawing' and how the properties of the polymer are altered by the drawing process.

EXPERIMENT 1.7   DRAWING A SEMI-CRYSTALLINE POLYMER

If you buy a four pack of drinks cans then there is a section of polymer webbing that holds the tops of the cans together. This polymer is semi-crystalline polyethylene. If you pull one of the loops with your hands then the polymer is drawn resulting in a large amount of permanent deformation. You should notice how a stable necked region forms initially and then this necked region extends along the length of the polymer. When the necked region extends along the whole length you are pulling it becomes very difficult to extend the polymer further. What is happening is that the polymeric chains in the amorphous region of the polyethylene are being aligned in the direction that you are pulling. To do this you only need to pull hard enough to break the Van der Waals bonds between the chains. As you cause one area of the polymer to become aligned, i.e. the formation of the stable neck, then the adjacent areas become aligned, i.e. the necked region extends along the polymer length. Once the polymer chains are aligned to achieve any further extension requires breaking of the strong covalent bonds between the carbon atoms in the chain and hence is much more difficult. By drawing a polymer there is an increase in strength and stiffness of the polymer, although extendibility decreases. Hence many modern textile fibres are drawn, also nylon fibres e.g. fishing line.

1.2.3 Ceramics and glasses

Ceramics show strong covalent or ionic bonding between the atoms. This type of bonding is directional and ceramics tend to have crystalline structures.

For example carbon in the form of diamond shows strong directional covalent bonds. Alumina, Al_2O_3, has strong ionic bonds.

1.2.4 Composites

As composites are comprised of metals, ceramics and polymers the structure will be as for the components. However the properties tend to be influenced by the microstructure, see section 1.3.

1.3   Microstructures

Microstructures are considered on the scale of 1 to 1000 μm, i.e. not single atoms anymore but how the groups of atoms are arranged and how they affect the material. So in this section we need to consider the typical microstructures seen for the different material types:

1.3.1 Metals

The simplest situation is for a single phase material, where a phase is a portion of the material having distinct chemical composition or/and physical structure. If a material is single phase (for example any pure elemental metals or alloys such as α-brass, low carbon steel) then it can be single crystal or polycrystalline. If it is single crystal then it is not until you start resolving the atomic positions that you can see the structure of the material. Polycrystalline materials are comprised of grains. A grain is a region where the crystal structure is continuous, i.e. a single crystal, and grain boundaries are formed where one grain meets its neighbour because the orientation of the crystal lattice is different. For a single phase material the microstructure could be comprised of equiaxed grains such as shown in figure 1.7 (equiaxed means that the grains are very approximately spherical). Typical structural alloys such as steels have grain sizes in the range of 5 - 30 μm.

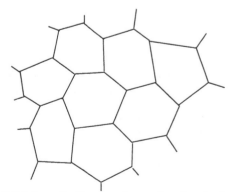

Figure 1.7 Schematic diagram of an equiaxed grain structure in a single phase material.

EXPERIMENT 1.8 EVERYDAY OBSERVATION OF GRAIN STRUCTURES

Is it possible to see the grain structure of materials? - yes in some situations. Where components have been galvanised (zinc plated) it is often possible to see the grain structure with the naked eye – could obtain examples possibly from **DIY** stores of galvanised wall brackets or tell students to look on their way home from school as it is often possible to find galvanised lamp posts, Figure 1.8 or railings.

Figure 1.8 Photograph of a galvanised lamp post showing grain structure.

In general the grain size is much smaller – of the order of a few tens of micrometres. The grains also need not be equiaxed, for example if a material has been rolled or extruded then an elongated grain structure may be formed, figure 1.9.

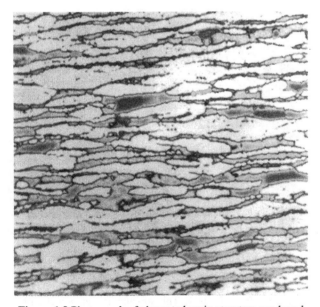

Figure 1.9 Photograph of elongated grain structure produced during rolling process.

EXPERIMENT 1.9 MEASUREMENT OF GRAIN SIZE

We often need to measure the grain size of a material (important because grain size can be altered by, for example, heat treatment and/or processing and affects mechanical properties) and hence we require a repeatable method. For equiaxed grain structures we can use the linear intercept method. Take an image of the microstructure e.g. one of the supplied photographs, and (using tracing paper or a photocopy of the photograph) draw a series of random lines across the structure. Measure the length of the lines, L, and count the number of times a grain boundary intercepts the line, N. The average grain size, D, of the material is then given by $D = {}^{3L}/_{2N}$ where the factor $^3/_2$ is there to account for spherical grains. Note that the measurement of line length has to be calibrated to the magnification of the photograph. An example is given in Figure 1.10.

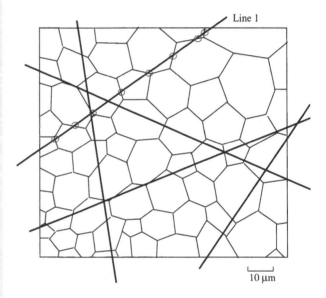

Figure 1.10 Schematic diagram of a grain structure showing methodology for measuring grain size using the linear intercept method. The intercepts between line and grain boundary are marked for one measurement line.

For line 1, in Figure 1.10, N = 8 and L = 79 μm. Hence D = 15 μm (for accuracy more lines need to be measured).

If this method is then considered for the elongated grain structure we can see that the results would not be meaningful hence another measurement is required. One technique is to take lines parallel and perpendicular to the elongated grains and determine the average lengths then to quote one of these and the aspect ratio i.e. (the length parallel to grains)/(length perpendicular to grains).

Does a single phase material have to show a grain structure?

No – it is possible to have what is called a dendritic structure. Many crystals grow, from a melt, in the first instance by branches developing in certain directions from nuclei. Secondary branches are later thrown out at periodic intervals by the main branches and in this way a skeleton crystal, or dendrite, is formed. The gaps between the branches are finally filled with solid which in a pure material is indistinguishable from the skeleton. Figure 1.11 shows a schematic diagram of a dendrite in two dimensions. (See Video).

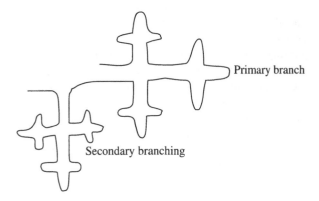

Figure 1.11. Schematic diagram of a dendrite.

What if the material is not single phase?
There are four different possibilities in this case:

1. *A mix of equiaxed grains of two distinct phases, for example a duplex stainless steel, Figure 1.12. Brass alloys are another example and consist of α- and β-phases.*

2. *A system where precipitation has occurred - i.e. precipitates or small particles have formed within a previously single phase material. For example in Al–Cu alloys $CuAl_2$ precipitates may be formed using a suitable heat treatment, Figure 1.13.*

3. *A system where the first phase formed from the liquid forms as dendrites with the second phase forming in the gaps between the dendrites, Figure 1.14.*

4. *An intimate mix of two or more phases other than as equiaxed grains. There are many complicated microstructures possible, examples are shown in figures 1.15 and 1.16.*

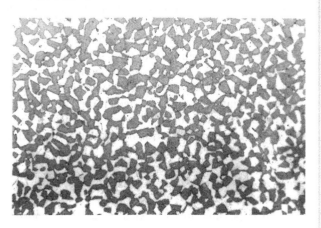

Figure 1.12 Photograph of duplex stainless steel showing a two phase material. ⊢ 25 μm ⊣

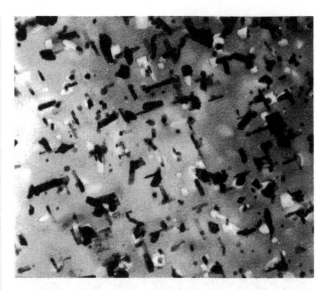

Figure 1.13 Photograph of $CuAl_2$ precipitates in an Al-Cu alloy. ⊢ 1 μm ⊣

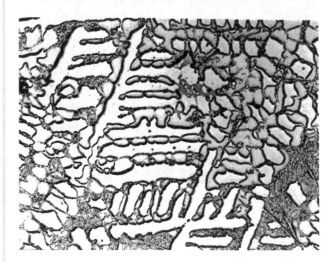

Figure 1.14 Photograph of dendrites in a two phase system. ⊢ 25 μm ⊣

Figure 1.15 Photograph of more complex microstructure - coarse pearlite, which is an intergrowth of two phases [lamellae of ferrite (body centred iron) and cementite (Fe_3C) in steel]. This structure is seen in high carbon (≈ 0.8 weight %C) steel. ⊢ 25 μm ⊣

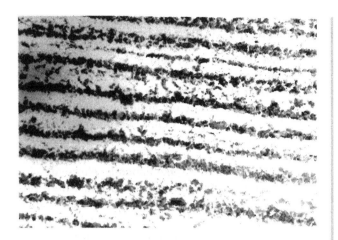

Figure 1.16 Photograph of more complex microstructure – rolled medium carbon (≈ 0.2 weight %C) steel showing bands of ferrite and fine pearlite.

$$\overline{\quad 10\ \mu m \quad}$$

1.3.2 Polymers

As mentioned earlier polymers can be amorphous or semi-crystalline. On a purely optical microscopy scale the microstructure of many polymers do not show distinct characteristics. An exception to this is for semi-crystalline polymers which crystallise by the formation of spherulites. Spherulites nucleate and then grow in a spherical manner – they are comprised of lamellae of folded polymer chains which grow out from the nucleation site. Figure 1.17 shows a spherulite in polyethylene viewed between crossed polars. The area between the spherulites is usually amorphous. (See Video).

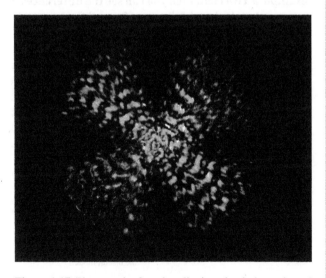

Figure 1.17 Photograph of a spherulite in polyethylene viewed between crossed polars.

Chain alignment (also called orientation) can also be produced by processing of polymers. For example during the injection moulding of plastic (polystyrene) rulers, set squares and protractors the pressure used to force the polymer into the mould, along with other factors, causes alignment of the polymer chains. This alignment makes the polystyrene birefringent (i.e. has different refractive indices in different directions) and hence, when viewed between crossed polars, shows fringes.

EXPERIMENT 1.10 OBSERVATION OF POLYMER CHAIN ALIGNMENT

View clear plastic (polystyrene) rulers, set squares and/ or protractors between crossed polars to see the chain alignment caused by the injection moulding process, figure 1.18. (This can also be used to determine the injection moulding entry point - by following the fringes back to the entry point - see chapter 3, section 3.2). Also refer to Experiment 1.7 on causing chain alignment through drawing.

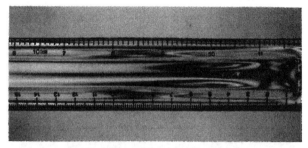

Figure 1.18 Photograph of fringes in plastic (polystyrene) ruler, viewed between crossed polars.

1.3.3 Ceramics and Glasses

Engineering ceramics are usually produced by sintering. This is the process of coalescing or fusing together of small particles to form a larger mass. High temperatures are used as diffusion is required. As ceramics are produced from powders the resulting microstructure is one of a polycrystalline material showing a grain structure. It is also often possible to see porosity in the material i.e. voids caused by the incomplete densification of the powder during sintering. Glasses are amorphous and don't show, optically, any discernible microstructure.

1.3.4 Composites

Composites are a wide range of materials which have very distinct microstructures. As composites are a mixture of two or more materials it is often easy to see its structure. However, it should also be noted that there is a difference between a composite structure and a composite material, although the distinction is often unclear, the examples given below are only guidelines.

Composite structures – padded envelope - brown paper and bubble wrap; bike helmet - expanded polystyrene foam and hard outer casing; mars bar - chocolate, nougat and caramel; chair - plastic, metal, foam, textiles; Steel reinforced concrete; Car tyres - rubber with steel reinforcing wires - try cutting (an old) one up!

Composite materials – many bike frames – aluminium + silicon carbide particles; tennis/squash racquet – carbon/graphite fibre reinforced epoxy; concrete - cement and sand/gravel

EXPERIMENT 1.11 COMPOSITE STRUCTURES OR COMPOSITE MATERIALS?

Ask students to find examples of composite structures (an integral component macroscopically of two or more materials) and composite materials (a material comprised microscopically of two or more materials and used in components).

What does the microstructure of composites look like? It depends whether the composite is comprised of particles, short or long fibres in a matrix. *Figure 1.19 gives some examples.*

a)

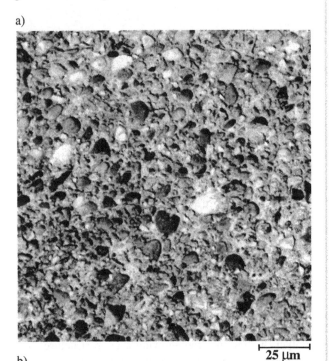

25 μm

b)

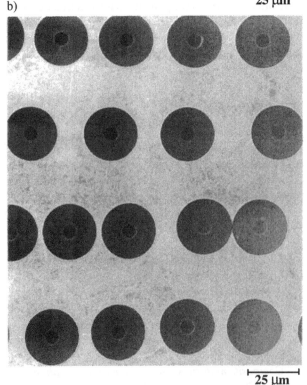

25 μm

Figure 1.19 Photographs of composite structures. a) concrete (composite of cement and aggregate) and b) SiC fibre reinforced aluminium showing the fibres from the end on.

EXPERIMENT 1.12 COMPOSITE STRENGTH AND FAILURE MODES

The idea in this experiment is to demonstrate the different types of reinforcement that can be used and to show their effect on fracture strength and type. The example here uses biscuits although other materials would work equally as well. Use a shortbread type mixture with liquorice boot laces as fibre reinforcement, both parallel and perpendicular to the way in which you break the biscuit, flake almonds, chocolate chip (particulate reinforcement) and plain. Make into rectangular slab biscuits then test by breaking them between your fingers - this can lead into discussion of failure modes and reasons why we use different types of reinforcement. For a greater degree of rigour you could use a three point or cantilever bending test and measure fracture loads (see Chapter 2, section 2.1) Note: Fibre reinforcement is very directional but you can see fibre pull out and crack bridging making it quite effective in increasing toughness, particulate reinforcement is less effective in these circumstances but is non directional. An example recipe for the biscuit mixture is given in the Appendix.

Other types of composite structures can be produced, for example ice and paper layered composite - here you can test the ice and the composite separately by hitting with a hammer to show the difference in energy absorbed (see experiment 2.11). Some chocolate bars are composites, for example Mars bars have layers of chocolate, caramel and a type of fudge. If you break a Mars bar and compare with a purely chocolate bar, for example a Twirl bar, then you can see the differences in energy absorbed and failure surface of the composite compared to a single constituent.

Generally a reinforcement is used because the fibres etc. are stiffer and stronger than the matrix and therefore give the composite a greater strength and stiffness than the matrix alone whilst the matrix provides the toughness. Often the reinforcement will also act to change the crack path in the material during fracture and may improve the toughness, Figure 1.20.

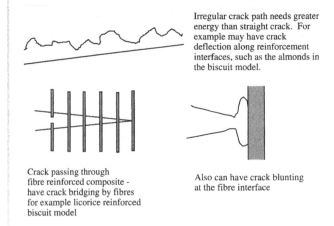

Irregular crack path needs greater energy than straight crack. For example may have crack deflection along reinforcement interfaces, such as the almonds in the biscuit model.

Crack passing through fibre reinforced composite - have crack bridging by fibres for example licorice reinforced biscuit model

Also can have crack blunting at the fibre interface

Figure 1.20 Mechanisms of crack growth in composites.

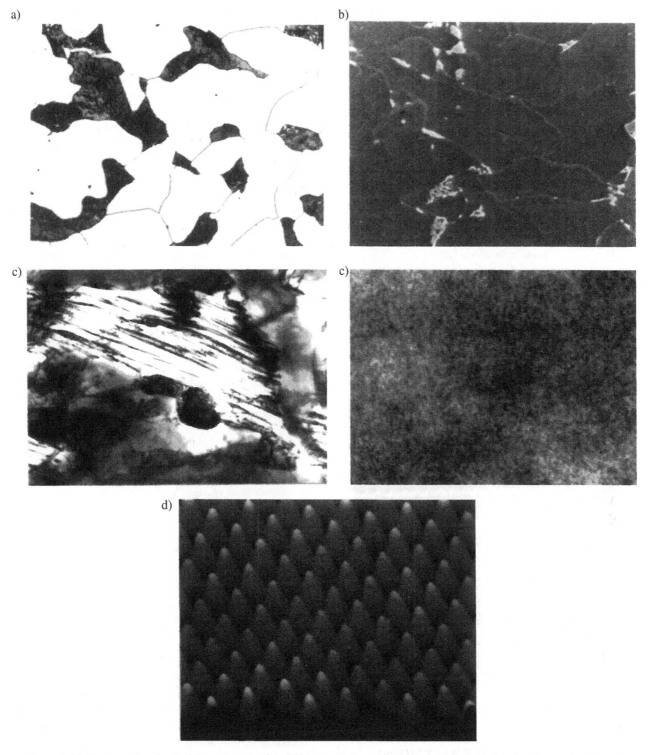

a)

b)

c)

c)

d)

Figure 1.21 Photographs of a steel (two phases) using different techniques: (a) optical microscopy (x100); (b) scanning electron microscopy (x300); (c) transmission electron microscopy (x1300); (d) high resolution electron microscopy (x50 000); and (e) atomic force microscopy (x400 000).

1.3.5 Observation of microstructure

In the previous sections microstructures have been discussed and shown using photographs produced by optical microscopy. There are other techniques that can be used to show the details of a microstructure. For example scanning electron microscopy, transmission electron microscopy, high resolution electron microscopy, atomic force microscopy. *Examples of images from these different techniques are given in Figure 1.21 - note the different magnifications.*

1.4 Crystal defects

There are a number of different crystal defects which occur in materials; dislocations, vacancies and solute atoms. A **dislocation** is a lattice imperfection in a crystal structure which exerts a profound effect on structure sensitive properties such as strength, hardness, ductility and toughness. A dislocation may have a configuration of an extra half plane of atoms inserted in the crystal structure, figure 1.22 and runs through the crystal, i.e.

is three dimensional although we can represent it in two dimensions. (Remember the description of a plane of atoms from figure 1.2.)

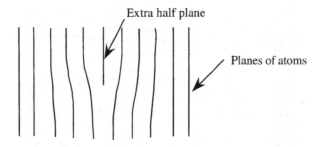

Figure 1.22. Schematic diagram of a dislocation - an extra half plane in the crystal lattice.

A **vacancy** is a site unoccupied by an atom or ion in a crystal lattice and a **solute** atom is an atom of a different species from that comprising the lattice, it may exhibit a size difference to the atoms of the lattice creating lattice strain.

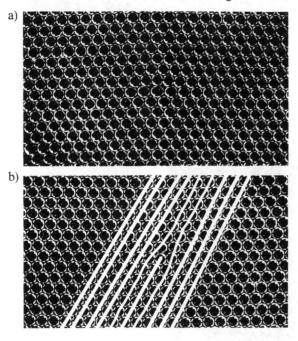

Figure 1.23 Photographs of (a) a bubble raft and (b) showing the dislocation.

EXPERIMENT 1.13 BUBBLE RAFT MODEL

Observation of dislocations, vacancies and solutes using a bubble raft (See Video):
Prepare a bubble raft in a plastic container containing a solution of washing up liquid and water. Use an air pump to form a few bubbles per second, note that the best results will be obtained with slow bubble rates and small (<1 mm diameter) bubbles. Gently sweep the bubbles towards one end of the tank - use a wooden baton or spring grip at either end of the tank. When a raft of size approximately 10 cm x 10 cm has been made, remove the air jet and carefully place the other baton or spring grip parallel to the first one so as to enclose the raft. Slight shearing motion of the grips will usually

enable you to rearrange the raft into a 'single crystal' i.e. bubbles in a perfect lattice with no grain boundaries. Place one of the batons or spring grips against the raft and move the other one gently in and out and sideways to simulate deformation by compression, tension and shear. Note that the movements are accommodated by elastic and plastic deformation.
(a) What is the mechanism of elastic deformation?
Bubbles change shape and move slightly apart as the rows try to move over one another.
(b) What is the mechanism of plastic deformation?
Dislocation motion
Figure 1.23 shows a reproduction of a photograph of a stationary dislocation in an otherwise perfect region of a bubble raft. Note that you could draw the lines in figure 1.23b) from top right to bottom left or top left to bottom right and still show the dislocation.

Burst a bubble, note what happens to the six bubbles around it - *the six surrounding 'atoms' all move inwards towards the vacant site, slightly.* **This vacant site is called a vacancy in the crystal structure, figure 1.24. If a bubble of a different size is introduced e.g. a larger bubble then there are slight distortions in the lattice around the foreign bubble. This represents the situation when a solute atom is present in the material i.e. a solid solution, e.g. Zn atoms in Cu-Zn (brass).**

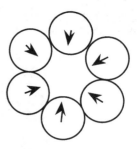

Figure 1.24 Schematic diagram of a vacancy.

Dislocations are produced when a metal sample has exceeded the limit of elasticity, i.e. when the material has yielded, and are required in order for the metal to show plastic deformation. As plastic deformation proceeds more dislocations are produced which can interact and become immobile thereby acting as obstacles to further dislocation motion. This is the process of work hardening. The extent to which a metal or alloy work hardens depends upon its crystal structure and composition, and has significant effects on the strength of the material. Figure 1.25 shows force displacement curves illustrating three different work hardening characteristics. In this case the material shows the same yield point but will have different tensile strengths (see Chapter 2 section 2.2 for definitions and explanation of terms).

If a metal is deformed plastically, e.g. by bending, stretching etc. then it will work harden. If the material is then tensile tested to determine its yield point it will be different than the non work hardened sample, see figure 1.26.

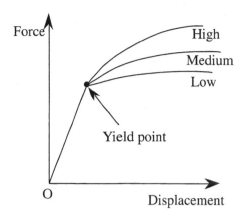

Figure 1.25 Force-displacement curve showing different work hardening characteristics.

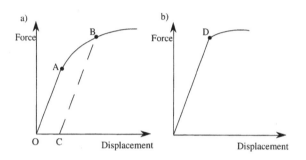

Figure 1.26 (a) Force displacement curve for a metal with high level of work hardening. **A** represents the yield point and **B** represents point to which the sample is loaded during an initial test. On removing the load the material follows line **BC** indicating a permanent deformation of **OC**. b) If the material is re tested then its yield point will be at **D(=B)** due to the work hardening effects.

The effect of work hardening can be removed by a heat treatment process known as annealing. Annealing a metal involves heating it to a sufficiently high temperature for the dislocations to move under thermal activation and annihilate each other. This then returns the material to the non work hardened condition and, in Figure 1.26, would give a yield point at **A**. An annealing heat treatment does not involve any phase transformations (see experiment 3.5 in Chapter 3).

EXPERIMENT 1.14 CONSEQUENCE OF DEFECTS

This experiment will look at the effect of dislocations on the strength of a metal. Copper strips ($\approx 100 \times 5 \times 2$ mm) can be used to show the effects of work hardening and annealing treatments. If possible relatively high purity copper should be used. Samples should be tested using cantilever bending or three point bending (see experiment 2.1 in Chapter 2) . A qualitative measure can be made by bending the samples by hand.
Strips should be heat treated to give the annealed condition. For accurate results a furnace should be used to give different annealed conditions; e.g. heat treatment

times and temperatures of 60 minutes at 350 °C, 500 °C and 800 °C can be used. Alternatively heat treating in a Bunsen burner flame can be used to produce one annealed condition, you will need to heat until the copper just starts to glow. Care must be taken in handling the annealed copper as any deformation (bending etc.) will work harden the sample and no differences will be observed. The as-received copper strips will normally be work hardened to some degree, bending them a little then straightening will be sufficient to make them work hardened.

On testing the samples in three point bending, or by hand, the annealed sample is much easier to bend (i.e. requires less force) than the work hardened sample. If the annealed sample is tested again after having been straightened then it will be much harder as it will have been work hardened. Figure 1.27 shows the results for strips tested under three point bending. The as received sample shows elastic deformation over the entire load range, the sample heat treated for 60 minutes at 350 °C yields at about 22 N and the other two samples yield at about 7 N. This is because the sample heat treated for 60 minutes at 350 °C has not been completely annealed whereas the other two samples have been.

The phenomenon of work hardening gives rise to a well known demonstration: take a bar of annealed copper and ask someone (a girl/women) to bend it into a horseshoe shape. Then ask someone else (a boy/man) to bend it back - you will find that it is almost impossible for them to do so. The initial annealed bar is relatively soft and easy to bend however the bending work hardens the bar making it much stronger and therefore more difficult to straighten. Note: the copper needs to be of a relatively high purity and should be annealed after the demonstration to return it to the soft condition.

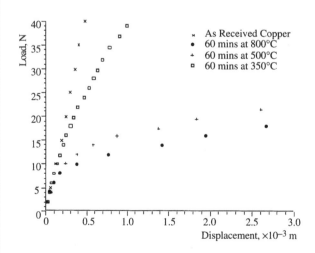

Figure 1.27 Force-displacement curve for as received and annealed copper strips.

APPENDIX

Recipe for the biscuits used in experiment 1.12

The biscuits need to be quite firm without being too brittle. This is a suggested recipe, numerous others would also work fine.

> 7 oz plain flour
> 1/4 teaspoon salt
> 1 oz semolina
> 4 oz margarine
> 4 oz castor sugar
> 1 beaten egg

1. Mix flour, salt and semolina

2. Rub in the margarine

3. Add sugar

4. Bind with the beaten egg to make firm paste

5. Roll our on floured board - make into require composites using additions. Place on a lightly greased baking tray.

6. Bake near the top of the oven at Gas Mark 5 (375°F) for ≈ 12 minutes until lightly golden. (Cooking time will vary depending on the composite being made and the thickness of the sample)

7. Leave for 1-2 minutes on tray to crispen then remove to cooling rack.

CHAPTER 2
MECHANICAL PROPERTIES
List of Experiments

MECHANICAL PROPERTIES

2.1 Stiffness

The stiffness of a material is its ability to resist elastic deformation.

EXPERIMENT 2.1 STIFFNESS OF MATERIALS

A simple experiment to determine the stiffness of various materials can be carried out using cantilever bending or three point bending: Note that in both cases it is important to plot the load vs. deflection curve as the experiment proceeds as you do not want to exceed the yield point of the material i.e. stop once the curve is no longer linear or if, on removing the weights, the material does not return to its original shape.

Ideally you want materials with the same cross-sectional area so that the differences are immediately obvious, however you can still carry out the experiment using different cross-sections. The balsa wood and beech square section rods can be obtained from most model shops. Some of the metal samples can be obtained from DIY stores or from specialists e.g. try the Goodfellows catalogue or RS Components (you can get their catalogue by writing to National Office, RS Components, PO Box 99, Corby, Northants, NN17 9RS tel. 01536 201201, fax. 01536 201501).

1. **Cantilever bending**: a simple experimental set-up is shown in figure 2.1.

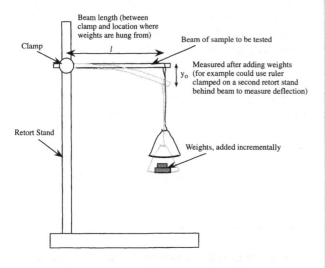

Figure 2.1 Experimental set up for cantilever bend testing to determine stiffness

Using beam theory the formula relating deflection to stiffness, measured by the Young's modulus, E, can be determined (see Appendix for optional mathematical derivation).

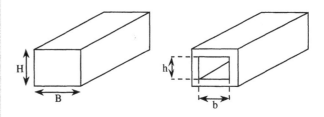

$$E = \frac{F}{y_o I} \frac{l^3}{3}$$

Where y_o is the deflection (mm), l the length between the fixed end and the load (mm), F is the load (N) and the second moment of area, I (mm⁴) is given below, see figure 2.2. E is then determined in MPa (1 MPa = 1 N/mm²) but is normally quoted in GPa (1 GPa = 10^3 MPa). You need to plot y_o against F, measure the gradient (F/y_o). From the knowledge of both l (fixed) and I (determined for each rod), E can be calculated. I is defined as:

$$I = \frac{BH^3}{12} \text{ for rectangular sections and } I = \frac{(BH^3 - bh^3)}{12}$$

for box sections

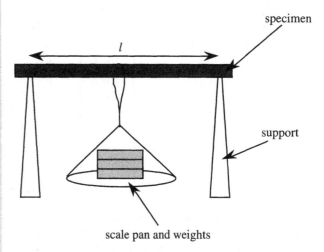

Figure 2.2 Specimen geometry

2. **Three point bend**: a simple experimental set-up is shown in figure 2.3. Three point bend tests can be carried out using simple equipment in the class room or purpose built rigs can be bought, for example from Plint and Partners (Oaklands Park, Wokingham, Berkshire, RG11 2FD).

Figure 2.3 Experimental set up for three point bending.

You can determine the relationship between stiffness and deflection for this arrangement (see Appendix - note this is provided for information only and the detailed mathematics can be ignored).

$$E = \frac{F\,l^3}{48\,I\,y_0}$$

Where E is the stiffness measured in terms of the Youngs Modulus (MPa), l is the distance between the two lower supports (mm), F is the load (N), y_0 is the displacement caused by the load (mm) and I is the second moment of area (mm⁴).

Some typical results for three point loading are shown below (note that the results tabulated show a degree of experimental scatter for the measurement of Youngs Modulus for brass). These values were calculated for the experimental conditions I used and will therefore differ from your results as sample cross sections etc. change.

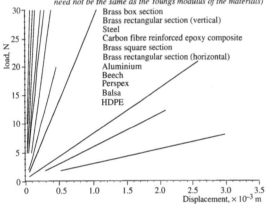

In order of load/displacement - *note that this order will vary if different cross sectional area samples are used in the expt. (i.e. this order need not be the same as the Youngs modulus of the materials)*

Brass box section
Brass rectangular section (vertical)
Steel
Carbon fibre reinforced epoxy composite
Brass square section
Brass rectangular section (horizontal)
Aluminium
Beech
Perspex
Balsa
HDPE

Figure 2.4 Results from three point bend tests

	E/GPa	ρ/Mgm⁻³
Steel	251.6	7.8
Composite	134.3	1.5
Brass, rect. sect. (Vert.)	99.0	8.5
Brass, rect. sect. (Horiz)	97.4	8.5
Brass, box section	101.8	8.5
Brass, square section	101.6	8.5
Aluminium	72.7	2.7
Beech	9.2	0.5
Balsa	2.0	0.1
Perspex	3.5	1.2
HDPE	1.2	0.96

Using this method it is not only possible to show the difference in stiffness between different materials but also to show how different sections influence stiffness and specific stiffness i.e. stiffness divided by density. This is important since very often we want to minimise the deflection and weight of a component e.g. bridge, aircraft wing

We can minimise y_0 by using   1. a short beam (i.e. minimise l), or
2. we want a high value of EI.

EI is known as the 'Flexural Rigidity'

Can either   1. increase E - materials selection
2. increase I - second moment of area, change by design i.e. geometry

E.g. I-beam - very common in structural materials, also tubes and corrugated structures

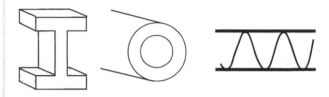

These structures have much greater values of I for the same volume of material than solid sections.

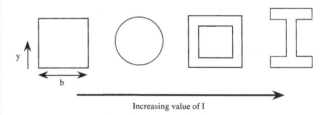

Increasing value of I

DATA HANDLING EXERCISE - stiffness of materials. Using the information below it is possible to determine the Youngs Modulus of a number of materials. The data presented was obtained using three point bending as described above in experiment 2.1.

I have left all the measurements in millimetres because the units cancel and it is easier to deal with the numbers in millimetres. It would be equally valid to carry out the calculations in meters if you should wish. I have also quoted the value of Youngs Modulus in GPa (i.e. 10³ MPa) rather than in MPa because this is convention in the subject, again you may wish to leave the values in MPa.

I have given the table of data for the load vs. extension plots from which the gradient F/y_0 can be determined. I have also given a table of data which includes the F/y_0 values if you do not wish your students to plot the graphs. **The procedure for determining the Youngs Modulus is as follows:**

1. Using the data below plot a graph of load, in N, vs. extension, in 10⁻² mm. (Please note that different loads were used for different materials because of their different responses)

2. Use the graph to calculate the gradient in N/mm - note you will need to take into account that the x-axis is in 10⁻² mm. This gives F/y_0, in N/mm, which is the stiffness of the beam NOT the material and therefore depends upon the second moment of area, I. This is why the different brass geometries give different values of F/y_0 e.g. the rectangular brass section in the horizontal and vertical orientations. The values of F/y_0 are given below for you to check your answers with.

Load, N,	Extension, x 10^{-2} mm										
	Brass, rect. sect.- horiz.	Brass, rect. sect.- vert.	Brass, square sect.	Brass, box	Steel	Alum-inium	Perspex	HDPE	Beech	Balsa	Com-posite
1							5				
2	4.3						16	51	6.0	31	2.5
3							31				
4	9.5						50	110	14	66	4.9
5		1.6	4.5		4.5	2.8	63				
6	14.0						75	192	21	100	7.0
7							84				
8	18.2							300	28	141	9.1
9							109				
10	23.0	3.1	8.2	2.2	9.2	5.2	116		34	167	11.8
12	27.8						136		41.5	203	13.2
14	32.0						161		48		
15		4.5	11.3		13.8	7.6			55.5		16.8
16	35.8						180		62.5		
18	40.5						214		71		
20	45.0	6.4	15.6	4.4	17.3	10.0	247		78		22.2
25		7.8	18.7		22.8	12.2			92		16.5
30		9.3	22.1	6.6	27.1	14.4			107		31.8

	Brass, rect. sect.- horiz.	Brass, rect. sect.- vert.	Brass, square sect.	Brass, box	Steel	Alum-inium	Perspex	HDPE	Beech	Balsa	Com-posite
F/y$_o$	44	32	125	455	110	208	7.9	3.0	28	5.9	94

3. By using the table below, which gives the dimensions of the beams, and the equations given below, the second moment of area, I, can be determined.

	B/mm	H/mm	b/mm	h/mm	I/mm^4
Brass, rect. sect. (Horiz)	6.32	2.36			6.923
Brass, rect. sect. (Vert.)	2.36	6.32			49.646
Brass, square section	3.84	3.88			18.692
Brass, box section	6.16	6.16	5	5	67.906
Steel	2.94	3.0			6.615
Aluminium	4.78	4.78			43.504
Perspex	3.76	4.78			34.221
HDPE	3.9	4.9			38.236
Beech	5	4.8			46.08
Balsa	4.8	4.8			44.237
Composite	2.0	4.0			10.667

$I = \dfrac{BH^3}{12}$ for rectangular sections and $I = \dfrac{(BH^3 - bh^3)}{12}$ for box sections

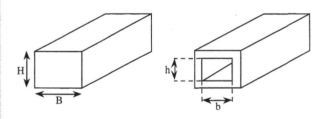

4. Now, using the values of I and F/y$_o$ that have been calculated, it is possible to determine the Youngs Modulus, E, of the materials. You will need to use the equation given below:

$$E = \frac{F \, l^3}{48 \, I \, y_o}$$

where l is 90 mm

21

Please note that E is quoted in GPa (= 10^3 MPa) in the table below because this is convention.

It should be noted that, within experimental error, the values of E for the different brass samples are now the same. This is because the Youngs Modulus is a material parameter and does not depend on geometry.

	B/mm	H/mm	b/mm	h/mm	I/mm⁴	F/y₀ N/mm	E/GPa
Brass, rect. sect. (Horiz)	6.32	2.36			6.923	44	97.4
Brass, rect. sect. (Vert.)	2.36	6.32			49.646	32	99.0
Brass, square section	3.84	3.88			18.692	125	101.6
Brass, box section	6.16	6.16	5	5	67.906	455	101.8
Steel	2.94	3.0			6.615	110	251.6
Aluminium	4.78	4.78			43.504	208	72.7
Perspex	3.76	4.78			34.221	7.9	3.5
HDPE	3.9	4.9			38.236	3.0	1.2
Beech	5	4.8			46.08	28	9.2
Balsa	4.8	4.8			44.237	5.9	2.0
Composite	2.0	4.0			10.667	94	134.3

2.2 Strength

The strength of many materials can be considered in several ways, three of which are - the yield strength, tensile strength and the fracture strength. The yield strength is the value of force divided by original cross sectional area of the sample when plastic deformation starts to occur, the tensile strength is the maximum force divided by the original cross sectional area of the sample, and the fracture strength is the value of force divided by final cross sectional area of the sample at failure, figure 2.5.

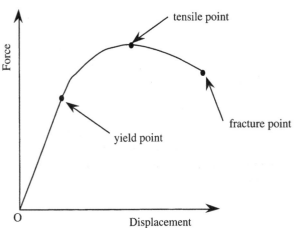

Figure 2.5 Typical force-displacement curve for many materials.

Determination of the yield point can be carried out using the experimental technique given in section 2.1. The tensile and fracture strengths are more difficult to determine without the more sophisticated equipment required to carry out a tensile test. (See Video). *Included in the pack are typical examples of the load-displacement traces for aluminium, perspex, polypropylene, mild steel and brass. You could use these graphs to allow students to determine the yield strength, tensile strength, and % elongation to failure, and hence compare materials.*

See also experiment 1.14 in Chapter 1 which demonstrates the effect of work hardening and annealing heat treatments on yield strength of copper samples.

EXPERIMENT 2.2 TENSILE TESTS

Use plasticene and sand to do a very simple tensile test manually - use a rolled cylinder of plasticene and pull gently, this will show excellent elongation and necking (necking is the localised reduction in cross sectional area) almost to a single point, especially if it is warm i.e. has been worked in the hands. However if you then use a sample that also has sand mixed in the plasticene you get void formation, see section 2.7, and reduced elongation/necking, figures 2.6 and 2.7. This simple experiment shows the effect of material purity on tensile properties. Can also relate to strain rate effects by pulling at different rates, see section 2.4 and the experiment using silly putty.

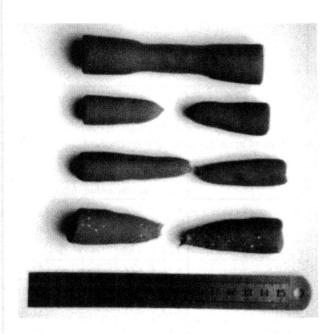

Figure 2.6 Photograph of simple tensile test samples using plasticene - from the top: standard tensile specimen shape, 'pure' plasticene showing necking to a point, plasticene with fine sand mixed in it showing reduced necking, plasticene with coarse sand mixed in it showing further reduced necking.

i.

ii.

150 μm

Figure 2.7 SEM photograph of tensile test surfaces,
.i. ductile fracture - note the deformation around the edge of
the fracture surface caused by the material behaving plastically,
and .ii. brittle fracture - very flat fracture, i.e. little/no plasticity.

For materials such as metals and polymers the value of,
for example, fracture strength that is obtained from a tensile
test or bend test will be relatively reproducible. For
materials that are quite ductile (plastic) in behaviour, i.e.

show good elongation (stretching) before breaking, this is
because the failure is due to plastic deformation, e.g.
necking. Even metals that behave in a brittle way, i.e. show
little ductility/plasticity, tend to fail from microstructural
defects which are reasonably consistent from sample to
sample. However for materials such as ceramics and
glasses the measured fracture stresses can show a large
amount of scatter. This is because failure originates from
surface or internal flaws such as scratches, porosity or
internal cracks and these are less consistent from sample
to sample, i.e. there is a large amount of variation in the
size of the defects that are present naturally in the material.
We can investigate the nature of the scatter by looking at
the fracture of glass. This can be done either as an
experiment or as a data handling exercise.

**EXPERIMENT 2.3 SCATTER IN FRACTURE
STRESS The experimental layout is shown in
figure 2.8.**

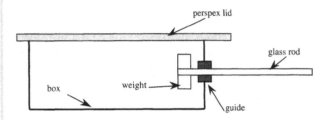

perspex lid

glass rod

box   weight   guide

Figure 2.8 Schematic diagram showing the experimental
layout for testing glass rods.

**You will need to place a *little* Vaseline on the inner
part of the guide to avoid introducing additional
scratches as the glass rod is pushed through. Glass
rods of ≈ 4 mm diameter and a weight of ≈ 5-7 kg
with a central hole of ≈ 4 mm diameter, to fit around
the glass rod, are required. Insert a rod through the
guide until *just* sufficient has emerged inside the box
for the weight to be placed on the rod so that it is
flush with the end of the rod. As soon as the weight
has been positioned, take your hand out of the box
and put the perspex cover in place. Keeping the rod
horizontal, gently but steadily push it into the box
until it breaks. Observe where the break occurs (is
it close to the part of the guide nearest to the weight?).
Carefully remove the broken length and measure the
distance, l', between the point of initiation of fracture and the end of the rod where the weight had
been. Subtract half the thickness of the weight to
obtain the distance, l, between the point of fracture
and the centre of mass of the weight. Record l, and
d, the diameter of the rod at the point of fracture. As
the glass rod often breaks uncleanly it is necessary
to clean the inside of the guide. Any shards of glass
will scratch the next piece of glass as you push it
through thereby introducing flaws which will ruin
your data. To clean the guide gently push through a
cotton bud and then re-Vaseline.**

Make as many separate measurements as possible of the distance l, in mm, defined above. Determine the maximum, tensile stress (σ_f in MPa) at fracture in each case from the formula:

$$\sigma_f = \frac{32Mgl}{\pi d^3}$$

where M is the mass of the weight in kg, d is the diameter of the rod at the point of fracture in mm, and g the acceleration due to gravity (9.81 ms^{-2}). Please note the units.

Abrade a number of lengths of rod by rotating them in coarse emery paper (grit size about 60 mm) and repeat the experiment and analysis as outlined above. It is then possible to look at the effects of abrasion on the mean strength and the level of scatter, e.g. standard deviation. A further test is to heat a further number of lengths of rod in a Bunsen flame, whilst wearing gloves!. Remember that glass softens as you heat it and will therefore sag if held horizontally supported only at one end. It is probably easiest to hold the rod vertically and move the Bunsen up and down the length. Allow to cool then hold the other end and repeat (be careful to put the hot rods on a heat proof mat to cool). This should fuse together any flaws present, it is obviously important to heat the glass rods evenly and thoroughly along their length so spend a little time doing this carefully. Calculate the mean fracture strength and scatter and compare with the earlier results.

DATA HANDLING EXERCISE -

The following results were obtained from a series of experiments using a weight of 7 kg.

A suitable data handling exercise for your students may be to give them the fracture length and rod diameter data and ask them to determine the fracture strength in each case, the average fracture strength and the standard deviation. Comments could be made on the accuracy to which the answer should be quoted and to the nature of the results.

Determine the fracture stress, σ_f in each case from the formula:

$$\sigma_f = \frac{32Mgl}{\pi d^3}$$

where M is the mass of the weight in kg, d is the diameter of the rod at the point of fracture in mm, l is the fracture length in mm, g the acceleration due to gravity (9.81 ms^{-2}) and σ_f is in MPa (MPa is the normal unit for stress and is equivalent to Nmm^{-2}). Please note the units used.

Condition	fracture length, mm	rod diameter, mm	σ_f, MPa
As received	12.0	3.98	133
	7.4	3.98	82
	10.7	3.98	118
	13.0	3.98	144
	11.8	3.98	130
	12.4	3.98	137
Abraded	6.7	4.1	68
	8.1	4.1	82
	10.2	4.1	103
	9.5	4.1	96
	8.6	4.02	92
	8.55	4.02	92
Heat treated	11.15	3.98	123
	11.4	3.98	126
	11.7	4.02	126
	13.2	4.02	142
	12.2	4.02	131

	Mean fracture stress, MPa	Standard deviation
As received	124	22.3
Abraded	89	12.4
Heat treated	129	8.4

The as received samples have good average fracture strength but show a large distribution of values. This is due to there being a range of surface flaw sizes being present from the manufacturing process and poor handling. The large standard deviation means it is necessary for any engineering application to be designed based on the average fracture strength divided by a large safety factor. It can be seen that the abrasion process introduces flaws that reduce the fracture strength of the glass rods but also narrow the distribution since failure will originate from the consistently large flaws present. The heat treated samples narrow the distribution appreciably, by blunting the largest flaws, and show good average strength.

2.3 Hardness

The hardness of a material is an important property that can be related to strength and is an easier quantity, than strength, to be measured. Hardness is a measure of a materials resistance to surface indentation or/and scratching.

EXPERIMENT 2.4 HARDNESS

The hardness of materials can be compared (or measured) using a simple test. You need a metal punch (obtainable from a DIY store), or a sharp hard nail (not as reproducible as the nail will blunt). Take a thick block of wood (ideally with a thickness about 10 mm less than the punch length) with a hole drilled through slightly bigger than the punch diameter. This block will keep the punch vertical. Fix a cylindrical guide over the hole - the length of the guide can be adjusted to give a reasonable size indent on the reference material but should be kept constant between tests. The radius of the cylinder should be slightly larger than the weight to be used e.g. a large steel ball bearing could be used with a 100 mm high cylinder. See figure 2.9.

Samples are tested in a reproducible manner by using the weight dropped from a constant height hence applying the same load on the punch for each test. The relative hardnesses of the materials can be compared by comparing the size of the indents. Ideally an optical microscope could be used to allow measurements of the indent size to be made.

Note that ceramics are very hard and you may not be able to make an indent on, for example, a piece of pottery. Instead the ceramic may crack showing its brittle nature. Indents on polymers need to be viewed immediately as some polymers can show time dependent properties and the indent may disappear.

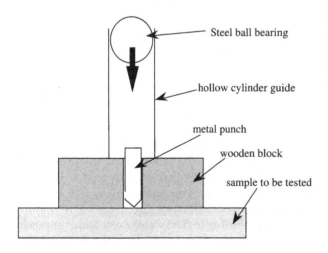

Figure 2.9 Schematic diagram of the arrangement for a simple hardness tester.

EXPERIMENT 2.5 SURFACE HARDNESS

Some components may have a hardened surface compared to the interior of the material. For example the more expensive candles are often composed of two waxes - an inner softer wax which looks dull and a shiny slightly harder outer wax. This is done as the outer wax burns more slowly so the molten wax is contained in a central well - i.e. a non drip candle. We can show the different hardness of the two layers by using temperature effects. Warm the candle. (This can be done by heating an oven to gas mark 1 for ≈ 5 minutes then switching off. Place the candle on a plate in the oven for ≈ 10 minutes or put the candle in hot - not boiling - water for 10 minutes). It should then be possible to bend the candle with ease. Immediately place the candle in cold water for about 10 seconds then try bending again. This time the surface layer will crack but the inner layer will bend. The outer layer is kept from falling off by the inner wax. The function of the hard outer layer is to protect the soft inner layer - and provide a more aethestically pleasing candle! We have chilled the surface layer such that it has become brittle. It should be possible to show this effect keeping the entire candle at the same temperature but practically this is hard to achieve.

There are many examples where materials with hard outer layers have been developed. For example steel with a hardened surface is used in the solid 'D' bicycle locks. The outer layer has been 'case hardened'. Case hardening is the term given to the hardening of the surface, or case layer. It can be achieved by various surface treatment processes to produce a different surface structure. The outer layer is so hard that it is very difficult to saw through it, however hard steel is also quite brittle so to stop thieves just hitting the lock and smashing it the core material is softer and more ductile.

2.4 Toughness

The toughness of a material is a measure of its resistance to fracture and can be measured by the energy absorbed during a fracture test such as a Charpy impact test. (See Video). *Another method of toughness testing is the Izod test, similar to the Charpy test. A miniature version can be bought from Plint and Partners (Oaklands Park, Wokingham, Berkshire, RG11 2FD). In both these cases the specimens tested are standardised and contain a blunt notch.*

EXPERIMENT 2.6 NOTCH SENSITIVITY

To determine the importance of notches or pre-existing cracks in a material take a sheet of paper and holding both ends pull - the paper has a high resistance to tearing, however if you now take a pair of scissors and make a small sharp cut in the centre of one free edge and now pull you see that the paper tears easily. You can experiment by cutting out different types of notches from the paper - for example if you cut a shallow semi-circular

notch then you find that the paper is still very resistant to tearing, it is really only a sharp crack that allows the paper to be torn easily. This is because paper is notch sensitive.

One of the most important aspects of the toughness of materials is that it can change with temperature. For example many materials are considered to be tough (i.e. absorb a lot of energy) and exhibit ductile failure at room temperature but are brittle (i.e. absorb little energy) and show cleavage type failures at low temperatures. Explanations and examples of ductile and brittle failures are given in section 2.7. This phenomenon gives rise to a ductile-brittle transition curve for materials such as steel and nylon, figure 2.10. Not all materials show this behaviour, for example copper shows tough, high energy absorption behaviour down to low temperatures.

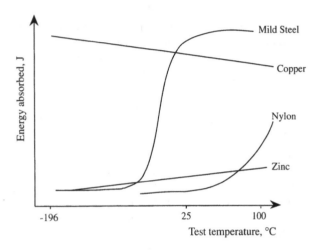

Figure 2.10 Schematic diagram of the energy absorbed for Charpy impact tests showing ductile - brittle transition behaviour.

EXPERIMENT 2.7 EFFECT OF TEMPERATURE ON TOUGHNESS I

If you take a material like chocolate, for example a finger from a Twirl bar etc., or toffee and warm it slightly (e.g. in your hands) then it is possible to bend the bar, if you continue to bend it, it breaks in a tearing manner, this is a ductile type fracture. Then if you take a finger of chocolate, or toffee, that has been in the refrigerator or freezer for a while and try and bend it, it is stronger but breaks quickly and without warning in a brittle manner like glass.

A similar experiment can be carried out with candles - take three candles of approx. 10-20 mm diameter and 150 mm length, cut off the wick. Place one candle in the freezer, leave one at room temperature and warm the third. (This can be done by heating an oven to gas mark 1 for ≈ 5 minutes then switching off. Place the candle on a plate in the oven for ≈ 10 minutes <u>or</u> put the candle in hot - not boiling - water for 10 minutes). Test the candles by bending and note the difference (if testing on a cold day then there may be little difference between the room temperature candle and that which has been

in the freezer). For the candle which has been warmed it should be possible to repeatedly bend through reasonably large deflections - if the temperature is too high then the candle may become squashy. This behaviour can be explained by the fact that paraffin wax is a polymer comprised of carbon chains of different lengths. At room temperature (25°C) all the chains are below their melting temperature, however at ≈ 50°C the shorter chains melt and are free to move - overall the wax candle is partially melted and softens. Note that to make the experiment reproducible you will need to take candles from the same box as different waxes melt at different temperatures.

Why is this behaviour important? Well many structures are made from steels and steels are one of the materials that shows this ductile to brittle transition. It is very important to avoid brittle failures as they are often catastrophic and can occur without warning. Some structures may experience a range of temperatures during operation. The most well known example is that of the Titanic which sank in the North Atlantic and this was due to the steel being brittle at those temperatures and the impact when the ship hit an iceberg. Different steels have different transition temperatures so provided the correct steel is selected then there are no problems in operation - such as oil rigs operating in the North Sea, pipe lines being laid across Alaska etc.

Probably one of the most well known examples in more recent history is the catastrophic failure of the Space Shuttle Challenger in 1986. The main contributory factor to the shuttle disaster was the failure of the O-ring seals on the solid rocket fuel booster. This allowed fuel to escape as a jet which ignited and impinged on the main fuel tanks causing them to explode. The reason for the failure of the O-rings was that on the day of the launch in February Florida experienced unusually cold weather leading to ice build up on the shuttle and surrounding building. The O-rings are made from a rubber which becomes less resilient below a certain temperature and therefore does not seal properly. (See Video).

EXPERIMENT 2.8 EFFECT OF TEMPERATURE ON TOUGHNESS II

It is possible to show the effect of temperature on the resilience of rubber to demonstrate the problem the Space Shuttle Challenger experienced using small rubber O-rings that you can buy from DIY stores. It can be difficult getting them to a low enough temperature using ice and water - you could try putting them in a freezer. The cold O-ring should take longer to return to its original shape after squeezing into an oval shape than the room temperature O-ring. Figure 2.11 shows some of the NASA images taken prior and during the flight illustrating the problem.

The same effect can be used by bouncing squash balls. Use three squash balls (of the same colour spot). Place one in the freezer, keep one at room temperature and warm the third in warm water. Measure the 'bounciness' of the balls by dropping them parallel to a metre ruler (from the same height) and seeing how high they bounce. The reason for the difference in behaviour is that the material in the warmer ball is acting like an elastic rubber. The cooler balls are approaching the 'glass transition temperature' (at around -30°C for the rubber). The glass transition temperature is the temperature below which the polymeric chains have lost their mobility. At this temperature the energy of the bounce is absorbed by the rubber and turned into heat energy. This is why squash players spend time warming up the squash balls before playing matches. Note that the different colour spot squash balls are made from slightly different rubbers and have slightly different glass transition temperatures.

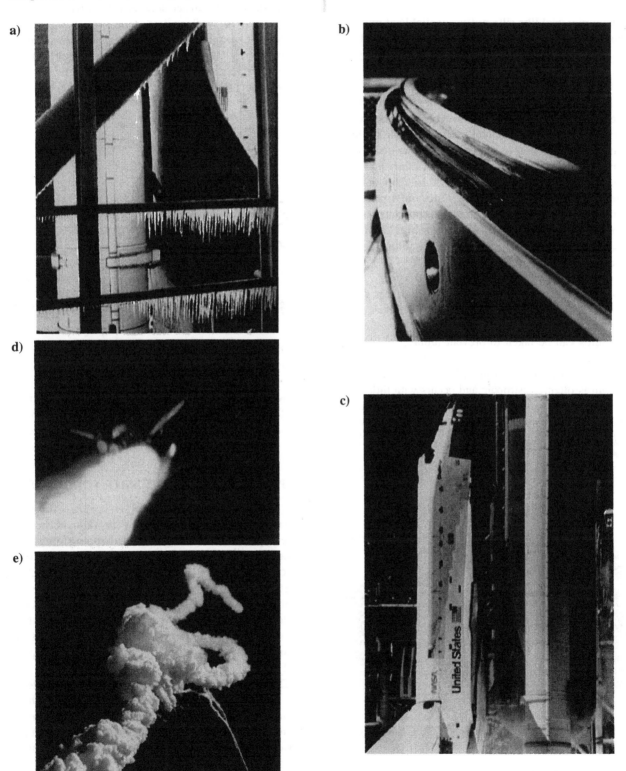

Figure 2.11 Photographs of the Space Shuttle disaster:
a) ice on the launch pad, b) O-ring seals, c) puff of brown smoke from the solid rocket-fuel booster (SRB) just prior to take off, d) jet of flame from SRB during flight, e) explosion of shuttle.

Another important variable with testing is strain rate i.e. how fast you deform the material. Some materials are strain rate sensitive - in other words they will behave differently depending on how fast you deform them.

EXPERIMENT 2.9 STRAIN RATE EFFECT ON TOUGHNESS

Take a material like silly putty - available at most toy shops - if you pull a piece in your hands relatively slowly then it extends until it has thinned almost to a point then it will break. However if you pull it very quickly (it helps if the silly putty isn't too warm for this) then you can get it to snap after a much more limited amount of extension. This type of behaviour is important when forming materials, for example the pressing of car body components whether in steel or aluminium needs to be carried out at a strain rate that will allow optimum productivity but avoid cracking. It also demonstrates the importance of testing a material under the conditions it experiences in an actual component as the material response can depend upon testing conditions such as strain rate (we have seen the importance of temperature and notches earlier).

Another important concept is fast and slow crack growth (not under cyclic loading which is fatigue). If a crack grows slowly under an applied load then it can be detected and measures can be taken to prevent failure, for example pressure vessels and pipe lines are designed to show slow crack growth so that they leak before breaking. Fast crack growth is often catastrophic and occurs with little or no warning. *For example ships constructed during the Second World War to replace convoy losses (the Liberty ships) suffered fast fracture and broke in half - literally - due to the construction technique used, figure 2.12.*

Figure 2.12 Photograph of Liberty ship S.S.Schenectady that has failed by fast fracture.

EXPERIMENT 2.10 FAST AND SLOW CRACK GROWTH

Using balloons it is possible to show fast and slow crack growth and crack arrest. If an inflated balloon is burst

with a pin then fast crack growth occurs and often the balloon ends up in two or more pieces. However if you place three pieces of cellotape on the inflated balloon to leave a small triangular area, figure 2.13, then burst the balloon in that triangle different behaviours can be observed. Initially there is fast crack growth however the crack is then arrested by the cellotape showing the existence of a stable crack. Generally (if done correctly) it is then possible to see slow crack growth below the cellotape, if the balloon lasts long enough (i.e. all the air doesn't escape) then when the crack grows through the reinforced area it will grow quickly again causing the balloon to pop. It may take a little practice to get it to work perfectly!

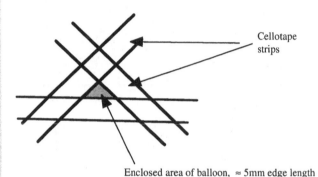

Cellotape strips

Enclosed area of balloon, ≈ 5mm edge length

Figure 2.13 Diagram showing arrangement of cellotape on balloon.

The arrangement of cellotape on the balloon can be thought of as a composite where the cellotape is being used to reinforce the rubber of the balloon against fast crack growth. We have seen how using different composite reinforcement affects strength and toughness in experiment 1.12. We can use another example to show this effect:

EXPERIMENT 2.11 COMPOSITE TOUGHNESS

Use two identical plastic containers approx. 100 mm x 100 mm x 400 mm. Half fill one with water, in the other place layers of completely soaked, dripping kitchen roll paper to about the same depth. Then top up with water. Place in the freezer - remember it will probably take at least 24 hours to freeze. When frozen remove from the containers and place on the edge of a bench. Place a padded plank (e.g. wooden plank wrapped in cloth) on top and secure to the bench using a 'G' clamp. Hit both with a hammer. The pure ice will shatter, the composite will not.

The reason for this difference is that ice is brittle and the cracks, once initiated run easily through the material. This is similar to the behaviour of ceramic materials, cracks are usually hard to initiate because a ceramic is hard, however cracks do not become blunted once formed. In the ice composite structure the cracks are deflected and blunted by the paper sheets thereby toughening the material.

2.5 Fatigue

The process of fatigue is progressive slow crack growth, and eventual failure, under cyclic loading - it can occur even in an initially unflawed material. Fatigue crack growth then leads to failure of the component or structure after a long period of use. In aircraft most structural failures are due to fatigue, usually associated with vibration or pressurisation cycles. *For example within two years of introduction in May 1952, 2 Comet aircraft had been lost out of a fleet of 9 - due to fatigue failure which initiated from riveted panels close to poorly designed square windows, figure 2.14.*

Figure 2.14 Photograph of comet aircraft showing square windows
(photograph courtesy of British Aerospace plc.).

EXPERIMENT 2.12 FATIGUE FAILURE

Fatigue failure can be demonstrated by taking a paper clip and bending it many times, eventually it fails. Note that it takes several cycles of bending before the paper clip breaks, you can also show the effect of the severity of loading by bending different paper clips though different angles - you should see a difference in the number of cycles before the paper clip breaks. It is possible to see the damage that the loading cycles cause to the paper clip - for example if you run your finger nail over the paper clip the surface feels smooth, however if you repeat this after several loading cycles then you can feel the greater surface roughness. This can be shown using scanning electron microscope (SEM) images, figure 2.15. See also section 2.7 on fractography, in particular figure 2.19.

EXPERIMENT 2.13 FATIGUE CRACK GROWTH

Whilst the paper clip shows failure after a number of loading cycles it does not show the crack growing. A simple experiment that does show slow crack growth is to take a thin sheet of aluminium and cut a sample of approx. 40x20 mm and make a small cut in the middle of one of the longer edges then gently flex repeatedly - generally possible to stop periodically and view the crack growth. It is possible to use the aluminium from a drinks can (with care!).

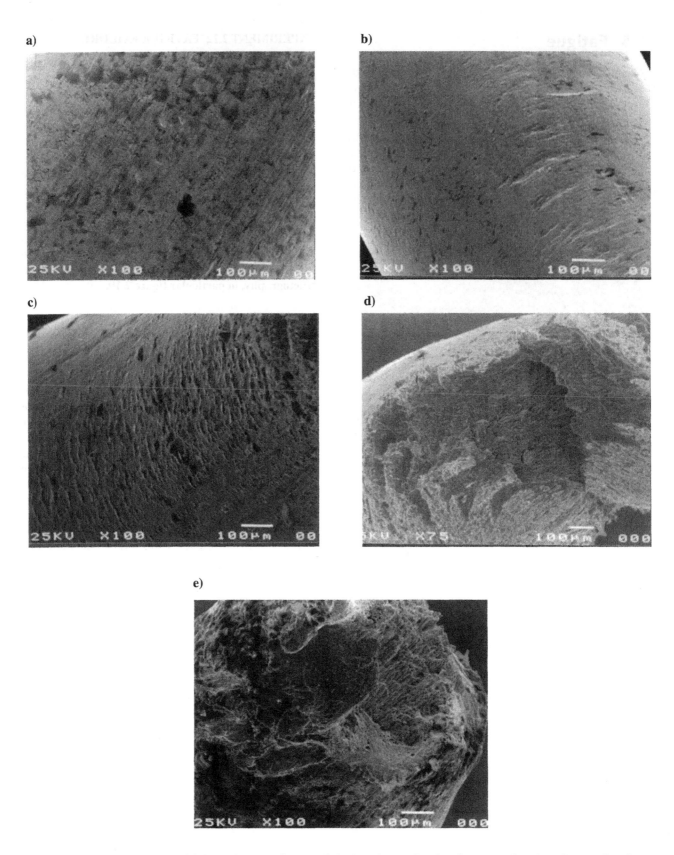

Figure 2.15 SEM photographs of fatigue of a paper clip: a) relatively undamaged surface from curved section of paper clip prior to fatiguing (notice a few flaws from production of the paper clip), b) area showing some degree of damage after initial couple of fatigue cycles, c) area showing further fatigue damage, d) partial failure of paper clip, e) final failure.

2.6 Creep

Creep is the process where a material shows a continuously increasing amount of deformation with time under a constant load. In practice creep is only important for temperatures $T > 0.3\,T_m$ for pure metals and $T > 0.4\,T_m$ for alloys and ceramics where T_m is the melting temperature of the material measured in Kelvin. Creep can be important in polymers at room temperature due to their low melting (or degradation) temperature.

Examples of where creep is important is in turbine blades in aero-engines, soldered joints in electronic components, lead pipes on buildings etc. Ice undergoes creep in glaciers.

EXPERIMENT 2.14 CREEP

It is possible to show creep using silly putty - if the silly putty is rolled into a cylinder and suspended, held, then it will creep under its own weight. Note that if you hold the silly putty in your hands for a while to warm it, it will creep more quickly - or vice versa if you put it in a refrigerator. You can also demonstrate creep using solder and a hair dryer to warm the solder - need to put weights on a coil of solder and watch it extend in length - can alter the temperature (i.e. use a hair dryer) or the load to examine the effects of load and temperature on the speed of the creep process, i.e. creep rate. The typical melting point of lead-tin solders is ≈183 °C hence room temperature is about 0.64 of the melting temperature. You can obtain a 500g reel of solder (e.g. Fry's Powerflow) at B&Q for around £4-5.

2.7 Fractography

Fracture occurs when a crack propagates across a sample. The initial crack may have been present before the sample was loaded or may have formed under load. Crack propagation may be a brittle process as in the fracture of glass, and also the cleavage of mild steel at low temperatures. Under such conditions, little work has to be done to propagate the crack. Examples of brittle failure are given in figure 2.16.

Alternatively, crack propagation can be a ductile process, involving the coalescence of voids, which requires large amounts of plastic deformation and tearing. This latter mode of fracture produces a dull fibrous fracture surface and leads to a high work of fracture, i.e. a large amount of energy is absorbed to break the material, figure 2.17.

EXPERIMENT 2.15 DUCTILE FAILURE SURFACES

Referring back to section 2.2 it was suggested to use plasticene with and without sand to show tensile properties. This can also be used to show what is meant by a ductile type failure. If the broken samples are examined, especially the ones with coarser sand or grit

a)

150μm

b)

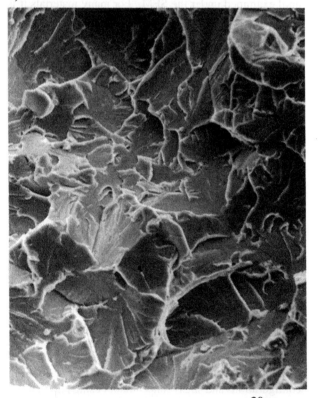

20μm

Figure 2.16 SEM photographs of brittle failure in a) zinc and b) steel. Note the difference in magnification, this is because the zinc sample had a very large grain size and cleavage occurs on one or more parallel planes in a single grain. For the steel sample the grain size is much smaller.

particles, then the surface shows little dimples in which lie grains of sand or grit. These dimples are known as voids and the same process occurs in many materials only on a finer scale. This process, during tensile testing, is shown schematically in figure 2.17 and photographs of typical examples of this ductile type failure are shown in figure 2.18.

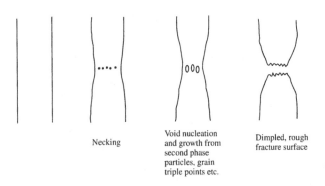

Figure 2.17 Schematic diagram showing the process of void formation during a tensile test.

Fatigue failures tend to be relatively flat and featureless until examined under the SEM. For many materials it is possible to see striations which are markings on the surface - generally one created for each loading cycle experienced by the sample, figure 2.19.

The fracture surfaces of composites will depend upon the type of reinforcement used. For example if fibre reinforcement is used then often it is possible to see fibres protruding from the fracture surface. However if it is a particulate composite then there is generally less evidence of the reinforcement on the fracture surface. A typical example of a fibre reinforced composite failure is given in figure 2.20.

2.8 Stress concentrations

A stress concentration arises when there are cracks, sharp corners, holes etc. in a component. They can cause premature failure of the component because the stresses in the material are greater in these regions. It is possible to determine the stress distribution in complex shaped components using computer software packages. It is also possible to examine the effect of different loading configurations and stress concentrators using photoelastic modelling. Here you use a birefringent material between crossed polars, apply a stress and examine the stress fringes produced. Note the difference in stress distribution for four and three point loading, figure 2.21.

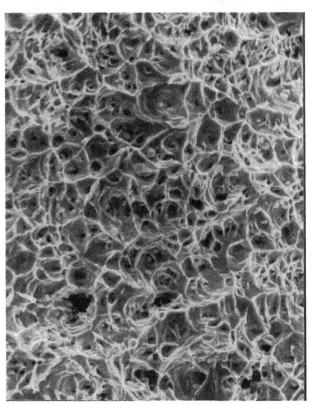

a)

20μm

b)

20μm

Figure 2.18 SEM photographs of ductile failure in a) copper and b) steel. The particles are oxide in copper and carbides in the steel. Note that between the main voids there are smaller voids and areas of ductile tearing to produce final failure.

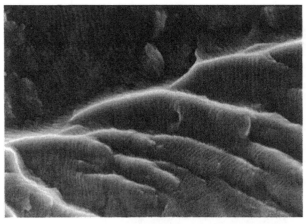

10μm

Figure 2.19 SEM photograph of the fatigue surface of an aluminium alloy.

500μm

Figure 2.20 SEM photographs of the fracture surface in a fibre reinforced composite.

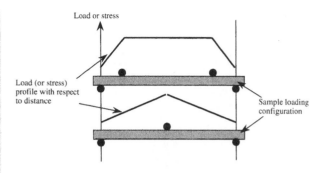

Figure 2.21 Stress distribution under four point (top) and three point (bottom) loading.

Figure 2.22 illustrates typical stress patterns for samples under three point and four point loading and containing various defects. Each different coloured fringe observed represents a contour of points at the same stress (similar to contours of height on OS maps). Where there are many fringes close together the stresses are high. Where there are few/no fringes there is little change in stress. The overall stress level and hence the number of fringes will depend upon the amount of loading on the sample which will be slightly different for the samples in figure 2.22. Note the large effect of a crack on the stress field.

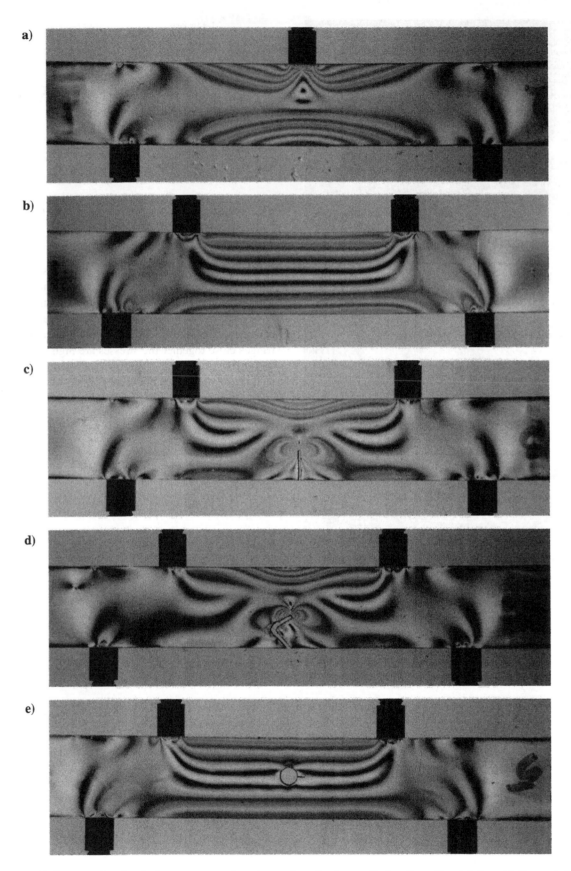

Figure 2.22 Photographs of typical stress patterns for: a) sample under three point bending without defect, b) four point loading without defect, c) four point loading with crack, d) four point loading with complex crack, and e) four point loading with central hole.

APPENDIX

Cantilever bending:

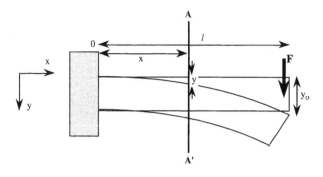

y$_0$ is the deflection of the end of the beam

Consider the beam cut at AA'

Force, F exerts a bending moment, M = F $(l - x)$

This must be balanced by the internal stresses in the beam.

Where $M = \dfrac{EI}{R}$

Curvature of the beam =

$$\frac{1}{R} = \frac{d^2y}{dx^2} = \frac{M}{EI} \quad \text{substitute for M}$$

Then $\quad \dfrac{d^2y}{dx^2} = \dfrac{F}{EI}(l-x) \quad$ equation defines the the beam

integrate $\quad \dfrac{dy}{dx} = \dfrac{F}{EI}\left(lx - \dfrac{x^2}{2}\right) + \text{constant}$

$\qquad\qquad$ constant = 0 as $^{dy}/_{dx} = 0$ at x = 0

integrate again $\quad y = \dfrac{F}{EI}\left(\dfrac{lx^2}{2} - \dfrac{x^3}{6}\right) + \text{constant}$

$\qquad\qquad$ constant = 0 as y = 0 at x = 0

Hence at x = l (end of the beam)

$$y_0 = \frac{F}{EI}\frac{l^3}{3}$$

Three Point Bending

For deflection of an elastic beam then:

Bending moment $\qquad M = \dfrac{EI}{R}$

where I is the second moment of inertia

$\qquad$ R is radius of curvature such that $\dfrac{1}{R} = \dfrac{d^2y}{dx^2}$

R can be calculated using the boundary conditions for the arrangement below

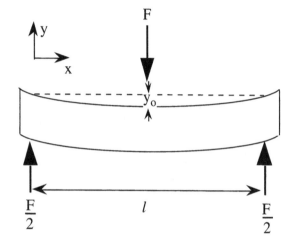

Boundary conditions are that: $\qquad$ x = 0, y = 0
$\qquad\qquad\qquad\qquad\qquad\qquad\qquad$ x = l, y = 0
$\qquad\qquad\qquad\qquad\qquad\qquad\qquad$ x = $^1/_2$, y = y$_0$
$\qquad\qquad\qquad\qquad\qquad\qquad\qquad$ x = $^1/_2$, $\dfrac{dy}{dx} = 0$

Using $\dfrac{1}{R} = \dfrac{d^2y}{dx^2}$ and

substituting into $M = \dfrac{EI}{R}$ and using $M = \dfrac{Fx}{2}$

where F is the applied load then:

$$E = \frac{F\,l^3}{48\,I\,y_0}$$

CHAPTER 3
PROCESSING
List of Experiments

CHAPTER 3
PROCESSING

3.1 Extrusion

Extrusion is the process of producing rods, tubes and various solid and hollow sections by forcing suitable material through a die by means of a ram. This technique is predominantly used for metals and plastics, for example medium density polyethylene gas pipes, unplasticised PVC water pipes, copper tubing, aluminium sections. In the case of metals the material is solid state extruded whereas plastics are extruded in the molten state. The extrusion process causes microstructural changes such as the alignment of polymeric chains or the elongation of grains in a metal.

EXPERIMENT 3.1 'EXTRUSION'

Take a large plastic syringe or equivalent and fill with alternate colours of plasticene. When the material is extruded out of the nozzle you can see the flow lines caused by the extrusion process, figure 3.1.

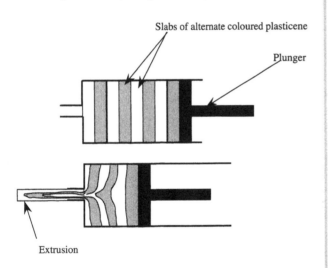

Figure 3.1 Diagram illustrating the extrusion processes using plasticene.

3.2 Injection moulding

Injection moulding is very similar to the extrusion process and is used extensively for moulding thermoplastics into complex shapes. It is the process whereby hot molten polymer is injected into a closed mould where it cools and solidifies. The mould comprises a cavity of the desired shape and can be split to remove the product, because of this it is possible to find 'witness marks' on the component indicating the location of the mould join.

Examples of injection moulded components are polystyrene CD cases, plastic rulers, set squares etc. The injection moulding process causes alignment of the polymer chains in, for example a plastic ruler, which can be observed in the form of stress fringes on viewing between crossed polars, see Chapter 1, section 1.3.2. It is also possible to determine the injection moulding entry point by tracing back the stress fringes.

EXPERIMENT 3.2 CHAIN ALIGNMENT DUE TO INJECTION MOULDING

As mentioned above it is possible to see the polymer chain alignment by viewing a component such as a plastic ruler between crossed polars. The chain alignment is forced by the process route, materials such as polystyrene are amorphous and therefore the chain alignment leaves residual stresses (frozen in stresses) in the ruler. It is possible, from the stress pattern, to determine the entry point of the mould as this is the point that the stress fringes radiate from, figure 3.2. You can show this using a simple experiment to observe the stresses, in the form of the stress fringes, between crossed polars. Then put the ruler in an oven at $\approx 70°C$ for half an hour. The increase in temperature gives the polymer chains sufficient energy to allow chain movement, the driving force being the frozen in stresses. The polymer chains recoil etc. relieving the stresses and on removing the ruler it can be seen to have shrunk! Also on viewing between crossed polars the stress fringes disappear as further evidence that the polymer chains have lost their alignment.

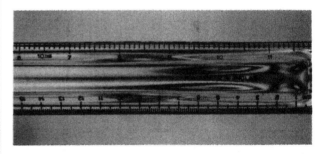

Figure 3.2 Stress fringes in a plastic (polystyrene) ruler - mould entry point can be seen on the middle of the right hand edge.

3.3 Casting

Casting is the process of pouring a molten material into a mould and allowing it to solidify. The material adopts the shape of the mould.

Examples of cast components are cast iron gates/cooking pans etc., Mg alloy helicopter gear boxes; Zn alloy toys, car door handles; Al alloy water pumps, motorcycle gear boxes. On a more domestic nature castings that we all come across are ice cubes, jellies etc.

EXPERIMENT 3.3 CASTING REQUIREMENTS

An important consideration in casting is that the fluid must fill the mould completely otherwise the mould shape will not be accurately reproduced. This is controlled by the fluidity of the material - casting alloys are designed to have high fluidity by controlling the composition. A simple experiment to demonstrate this is to 'cast' plaster of Paris figures but to make the plaster of Paris to different consistencies to see which gives the best casting. Economic considerations can also be introduced as for high productivity you want the casting to solidify as quickly as possible, i.e. requires a low fluidity. Can also look for defects such as porosity, inclusions etc. For example you can see porosity in ice cubes.

What about a measure of fluidity. In practice it is measured by the distance filled in coiled mould under set height of feeding, figure 3.3. It would be simple to set up a measure of fluidity and compare materials such as water, plaster of Paris, detergents, treacle etc. using coiled rubber tubing.

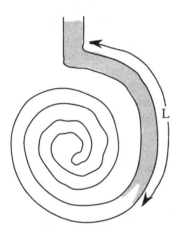

Fluidity - L
Greater the value of L, the greater the fluidity

Figure 3.3 Diagram showing a measure of fluidity

Casting, as with all processing techniques affects the microstructure of a material. For example with zinc ingot castings a very distinct grain structure is observed, figure 3.4. These microstructural changes are important in controlling the properties of materials.

3.4 Forming

Forming a material may be by pressing a shape from a sheet such as car body panels, or bending wire into coat hangers. One of the important considerations in forming materials is the strain rate used. This aspect has been discussed in Chapter 2, section 2.4.

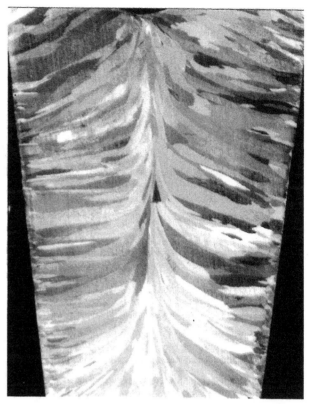

⊢——— 50 mm ———⊣

Figure 3.4 Grain structure in a zinc ingot casting

EXPERIMENT 3.4 STRAIN RATE EFFECTS IN FORMING

Use of silly putty to show effect of strain rate, see Chapter 2, section 2.4. (Experiment 2.9 on sheet 2.21)

3.5 Heat Treatment

Heat treatment processes are used extensively in the materials industry for changing a materials properties. Lets take the example of a carbon steel - one of the most important structural materials. There are a number of heat treatment techniques that can be used for steel which gives the material very different properties by changing its structure. We have mentioned structure of metals in Chapter 1 by considering the way in which atoms can pack together, for example face centred cubic or body centred cubic packing arrangements (see section 1.2.1). The different structures can have very different mechanical properties, for example strength, toughness etc. (see Chapter 2 for definitions and example). Steel can adopt different structures depending upon the heat treatment that it has experienced and hence the same steel sample can be made to have different properties.

Before we start with the details of an experiment it will help to go through a few definitions of heat treatment terms relevant to carbon steels and the structures that they can adopt:

Quenching The term quenching refers to when the steel is cooled very quickly from a high temperature, for example by plunging into cold water after being held at > ≈ 850°C. By quenching the steel a very strong (hard) and brittle structure (called martensite) is formed which has poor ductility.

Tempering Tempering is where a quenched piece of carbon steel is held at a temperature of ≈ 200°C, or higher, for a period of time, typically around 1-2 hours. A quenched and tempered piece of steel will have moderate to good strength but will be more tough and ductile than a purely quenched structure. The structure is known as tempered martensite.

Normalising The process of normalising involves holding the piece of steel at ≈ 850 °C, or above, then cooling slowly. This develops a structure known as ferrite and pearlite (see 'Photographs' accompanying pack, photograph 2) which has a low strength with good toughness and ductility.

- Ductility is a measure of the ability of a material to be deformed, e.g. bent/stretched, before failing.

- Toughness is a measure of the resistance to fracture of a material, e.g. energy absorbed on breaking.

- Strength is a measure of the stress (load/area) required to deform a material, e.g. in a tensile test.

- Hardness is a measure of a materials resistance to surface indentation and can be related to strength.

Carbon steel can adopt a number of different structures, as mentioned above. I will mention these structures for interest although it is not important to use these in the experiment. The high temperature equilibrium structure (exists at temperatures > ≈ 850°C) is 'austenite' which is a face centred cubic structure (f.c.c.). The low temperature equilibrium structure is a mixture of 'ferrite' and 'pearlite', ferrite is body centred cubic (b.c.c.) and pearlite is itself a mixture of ferrite and 'cementite' (cementite is Fe_3C) - see microstructure section (1.3.1) in Chapter 1. The structure that is formed by quenching the high temperature phase, austenite, is a non-equilibrium phase called 'martensite' which is a distorted body centred tetragonal structure (b.c.t.).

EXPERIMENT 3.5 HEAT TREATMENT OF STEEL

You need to get some 3 mm diameter piano wire, for example from a model shop - smaller diameter wire can also be used as can masonary nails.

Normalising The steel from the supplier (as-received) is normalised. Take a 100 mm length of wire and bend it through 90°. You should just be able to do this using your hands since the normalised steel has low strength and high ductility.

Quenching Take a 100 mm length of wire and heat the middle using a Bunsen burner until to glows bright orange. You need to heat a length of about 1 cm. The temperature is now about 850°C. As quickly as you can, quench the steel by dropping it in a beaker of water. Break the steel using your hands. It should be extremely brittle because of the high carbon content of the untempered martensite i.e. the quenched steel has low ductility.

Tempering Repeat the previous heat treatment and quench to produce a brittle wire, then reheat the wire until it just begins to glow a dull red. The temperature should now be approximately 600°C to 700°C. The high carbon martensite has now been tempered. Cool the steel - you can quench it as it will not alter the structure further. Try to bend the steel using your hands. It will be stronger than the normalised steel, but you should be able to introduce a 90° bend, demonstrating that tempering restores ductility to quenched steels. This is because the quenched and tempered steel has moderate to high strength and high ductility.

Hints You should be able to bend the steel using just your hands. However, if you're not feeling very strong, the demonstration works well with smaller diameter wire (1.5 mm). Alternatively, place the wire in a bench vice and bend it with taps from a hammer. Take care to prevent the steel from flying towards the class if it breaks!

If you're having problems cutting the piano wire, you can snap it by transforming the region to be broken to martensite. If you find that the tempered steel is weaker than the as-received steel, you probably made the steel too hot during the tempering heat treatment. Caution: ensure that all the heat treated steel has been fully immersed in water before you handle it without gloves. Hot steel burns.

See also experiment 1.14 in Chapter 1 which demonstrates the effect of work hardening and annealing heat treatments on yield strength of copper samples.

CHAPTER 4
MATERIALS SELECTION
List of Experiments

CHAPTER 4
MATERIALS SELECTION

Materials are selected for applications by matching the properties of the material to the service conditions required for the component. The first step in the selection process requires that we analyse the application to determine the most important characteristics that the material must possess. Should the materials be strong, ductile, transparent, conducting or insulating, stiff, tough, etc. In service will the component be subjected to the repeated application of a load (for example artificial hip joint), a high stress at elevated temperature (turbine blades in aero-engines), high thermal stresses (the space shuttle on re-entry), wear conditions (shoe soles)?

Often the main information that we require are the mechanical properties of the material which not only determine how well the material performs in service, but also determines the ease with which the material can be formed into a useful shape. We have looked at a number of these tests in Chapter 2.

EXPERIMENT 4.1 LIGHT WEIGHT CRASH HELMET FOR EGGS

This works best in the form of a light-hearted competition in the manner of the "Great Egg Race" and would be ideal for the end of term as it is fun but also instructive. This experiment should emphasise the selection of materials as well as the design aspect. Initially introduce the development of crash helmets for human heads. For example the use of polymer foams and load spreading shells to mitigate the effects of localised forces on the skull, or of excessive acceleration of the brain.

Materials needed for this experiment are any that you can find - bubble wrap, cardboard, expanded polystyrene, old egg boxes, foam, wire, drinking straws, aluminium foil, fabrics etc. You'll also need cellotape, scissors and a balance.

The idea is to give the students their own eggs, weigh the eggs, then allow anything from 20 minutes to an hour+ to design and build a crash helmet (encompassing all the egg). This is then weighed and the weight of the egg subtracted to give the crash helmet weight. The crash helmet should then be tested - for example dropped from an upstairs window / down the stair well etc. The winner is the design that survives and is the lightest (we would recommend about a 3-4 metre drop but it doesn't really matter). You can add modifications such as the crash helmet must float or be water proof, have a volume less than x (as a person would not walk around with a huge helmet on!) etc.

VIDEO TIMINGS

Contents	Context	Section	Times
Space Shuttle Challenger disaster	Effect of temperature on material properties	2.4	≈ 22 mins
Space Shuttle tiles	Properties of ceramics - thermal insulation	1.1.3	≈ 6 min
Crystal growth	Microstructural development	1.3.1+1.3.2	≈ 11 min
Bubble raft model	Model of dislocations and grains	1.4	≈ 4 min
Mechanical testing	Tensile and impact testing	2.2 + 2.4	≈ 14 min

Timings:

0.20	**Space Shuttle Challenger disaster** – introduction and assembly of shuttle craft. Launch to explosion of shuttle. (Note – temperatures of low 20s refers to °F not °C)
7.21	Gap (blue screen)
7.31	Post flight analysis and telemetry data pinpointing cause of accident.
21.32	Gap (blue screen)
21.42	Presidential commission findings
22.18	FINISH 10 seconds black screen
22.28	**Space shuttle tiles** – thermal properties and structure.
28.37	FINISH 10 seconds (black screen)
28.47	**Crystal growth** in a single phase material – planar → cellular → dendritic
33.23	Gap (blue screen)
33.28	Crystal growth in a single phase material – faceted growth
33.53	Gap (blue screen)
33.58	Crystal growth in a two phase material – eutectic growth
34.31	Gap (blue screen)
34.36	Crystal growth in a two phase material – eutectic growth past impurity particle
34.52	Gap (blue screen)
34.57	Crystal growth in a two phase material – eutectic growth showing effect of change in growth rate
35.37	Gap (blue screen)
35.42	Crystal growth in a two phase material – more complex eutectic structure
36.12	Gap (blue screen)
36.17	Spherulitic growth structures
39.23	FINISH 10 seconds (black screen)
39.33	**Bubble raft** production and how it relates to the grain and crystal structures
41.21	Gap (blue screen)
41.26	Dislocation appearance and motion
42.27	Gap (blue screen)
42.32	Dislocation interactions
43.18	Gap (blue screen)
43.23	Recrystallisation
43.48	FINISH 10 seconds (black screen)
43.58	**Tensile testing** of steels
49.30	Gap (blue screen)
49.35	Tensile testing of plastics
52.12	Gap (blue screen)
52.17	**Impact testing**
57.57	FINISH (black screen)

Explanatory notes to accompany video

1. **Space Shuttle Challenger disaster** *Copyright NASA* Case study on the importance of correct material selection and the effect of temperature on material properties.

In section 2.4 on toughness the Space Shuttle Challenger disaster was linked to the loss of resilience of rubber O-ring seals in the solid rocket fuel booster. The video shows NASA footage of the assembly of the Space Shuttle, followed by a detailed post flight analysis using remote camera images which enabled the cause of the failure to be identified. The video could be used as general information for the students to the importance of material selection and knowledge of material properties under different operating conditions. A possible exercise for the students would be to write a short (1 page) article in the style of a popular science journal, such as *New Scientist*, on the cause of the failure and the importance of materials science/engineering in avoid such problems.

2. **Space Shuttle tiles** *Copyright NASA* Ceramic material properties and general interest information on the space shuttle tiles.

Section 1.1.3 gives some examples of ceramic materials and where they are used. The Space Shuttle tiles are given as one example. This section of the video illustrates some of the properties of these ceramic tiles – extremely low density, good heat resistance and extremely low thermal conduction. Information is also given on how the tiles are fixed onto the shuttle.

3. **Crystal growth morphologies** Illustration of different types of crystal growth to compliment microstructures (no soundtrack).

Section 1.3 gives examples of different microstructures that can be seen in metallic and polymeric systems. Photographs are given to illustrate the typical microstructures. This video accompanies this section allowing the grow of a material to be viewed. Chemicals are used to represent the types of crystal growth morphologies that are seen in metals. A temperature gradient exists across the material causing growth to occur from left to right on the screen. The first two examples are for a single phase material, such as pure copper, pure aluminium etc. Planar, cellular and dendritic growth both in a non faceted and faceted manner are shown. The type of growth depends upon cooling rate for a material. It is not possible to see dendrites in a pure metal as the boundaries between dendrites do not show up. **Figure 1.14** shows dendrites in a two phase material – the light phase has the dendrite morphology. In pure materials grain structures are observed, as shown in **Figures 1.7 and 1.8.**

The following four video sections are for growth in a two phase material where solidification of the two distinct phases is occurring concurrently giving rise to a 'lamellae' structure. This type of structure is called a eutectic structure. There are examples of simple eutectic growth with the lamellae growing parallel to one another; eutectic growth past an impurity particle (for example a piece of dirt in a mould) – note that the number of lamellae change due to the external influence; eutectic growth where the growth rate is changed (through a change in cooling rate) – here note the change in thickness of the lamellae; complex eutectic growth where the lamellae are no longer parallel (this is due to a change in temperature gradient away from simply across the screen).

The final video section shows the growth of spherulites in a polymer (i.e. crystalline regions in an amorphous polymer). This is discussed in section 1.3.2.

4. **Bubble raft model** Illustration of the use of a bubble raft model to demonstrate dislocations, grain structures and recrystallisation (no soundtrack).

This video section can be used to link with experiment 1.13 in section 1.4 to help the students obtain the most from their bubble raft, or it can be used to help explain dislocations, grain structures etc. when construction of a bubble raft is impractical.

5. **Mechanical testing** *Copyright The Open University* Illustration of tensile testing and impact testing in a research/industrial lab.

This video section can be used in connection with section 2.2 on strength and 2.4 on toughness. Students are able to carry out their own strength and toughness assessment. However it is important that they are able to see how more accurate measurements are made. The video will provide information on testing techniques. (It may be possible to arrange for the students to visit the Metallurgy and Materials Department in a local university and carry out some of these tests themselves).

PHOTOGRAPHS

Included are some representative photographs of microstructures and fracture surfaces for use in conjunction with the experiments/general teaching. You may wish to photocopy these onto overhead transparencies or use photocopies for grain size measurements etc.

Photograph list:

1 Photograph of duplex stainless steel showing a two phase material.
2 Photograph of ferrite and pearlite - a two phase system.
3 Photograph of elongated grain structure in a rolled material [duplex stainless steel].
4 Photograph of dendrites in a two phase system.
5 Photograph of pearlite - lamellae of ferrite and cementite (Fe_3C) in steel.
6 SEM photograph of ductile failure in copper.
7 Photograph of the grain structure in a zinc ingot casting.

Suggested uses:

Photographs 1-3 could be used in conjunction with **Experiment 1.9** on measurement of grain size. These photographs could also be used to calculate the percentage of each phase present.

Photographs 1-5 are examples of some of the different microstructures that can be seen in metallic systems and can be used as representative photographs to go with the teaching of microstructures *(see Chapter 1, section 1.3.1)*.

Photograph 6 can be used in conjunction with **Experiment 2.15** on ductile failure in plasticene.

Photograph 7 can be used to illustrate a typical cast microstructure showing grain growth from the mould walls resulting in columnar grains *(see Chapter 3, section 3.3 on casting)*.

1. Photograph of duplex stainless steel showing a two phase material

10μm

2. Photograph of ferrite and pearlite – a two phase system

5μm

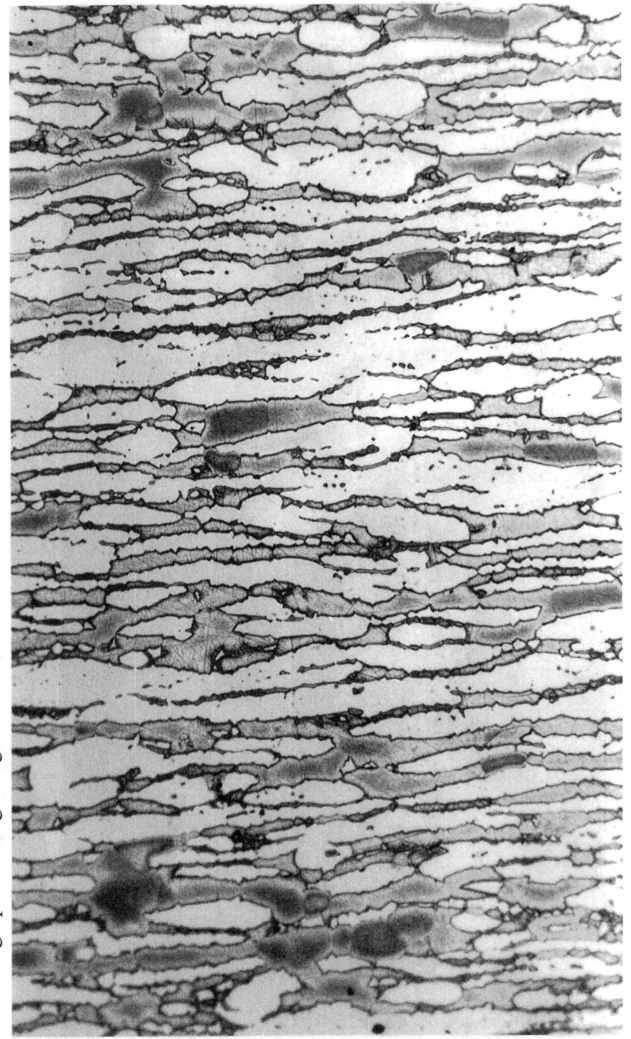

3. Photograph of elongated grain structure in a rolled material

20μm

4. Photograph of dendrites in a two phase system

10μm

5. Photograph of pearlite – lamellae of ferrite and cementite (Fe$_3$C) in steel.

10μm

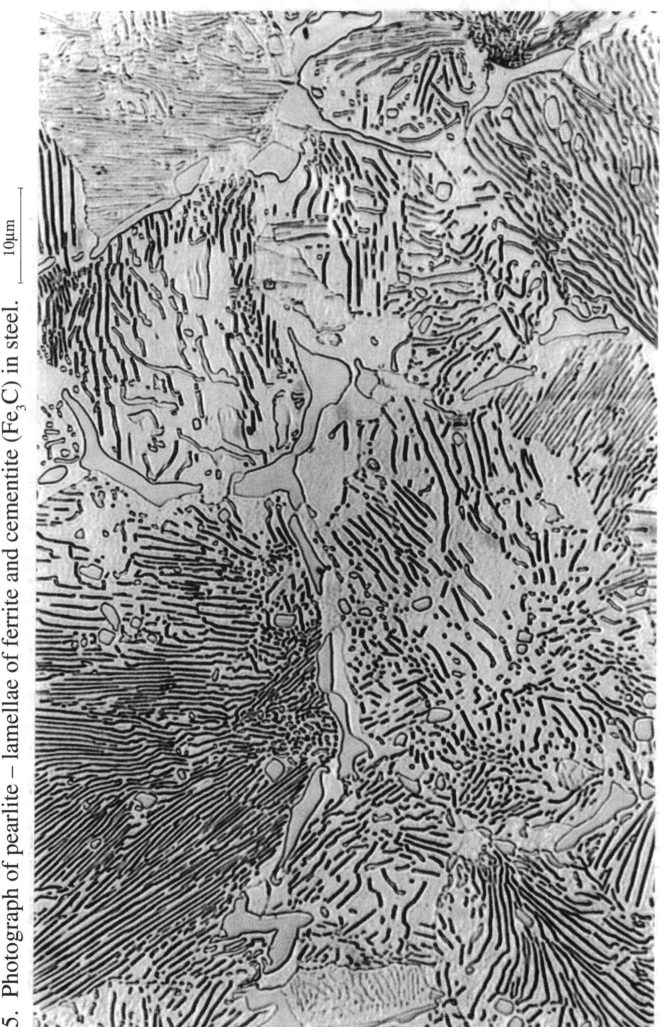

6. SEM photograph of ductile failure in copper

10μm

7. Photograph of the grain structure in a zinc ingot casting.

0.4 mm

TENSILE TEST DATA

Included are typical load-displacement curves for aluminium, PMMA (perspex), polypropylene, mild steel and brass. These can be used by students to calculate yield strengths, tensile strengths, elongation values etc. The sample measurements (thickness, width etc.) are given below. Please note that different chart speeds and load scales were used. The samples used were flat strip samples:

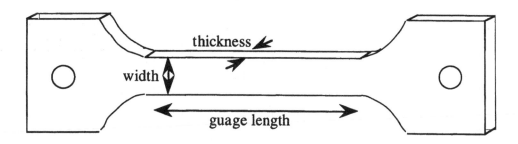

Material	Width, mm	Thickness mm	Cross sectional area, mm²	Original guage length, mm	Final guage length, mm	% Elongation	Yield Load, N	Max. Load, N
PMMA (Perspex)	12.75	3.03	38.63	65.0	65.10	0.15	1260	3070
Polypropylene	12.72	2.98	37.78	25.0	186.0	644	970	1230
Aluminium	13.08	1.23	16.09	65.0	84.6	30	300	1470
Brass	13.05	1.01	13.18	65.0	102.6	58	1080	4100
Mild Steel	13.01	0.88	11.45	65.0	89.0	38	2300	3440

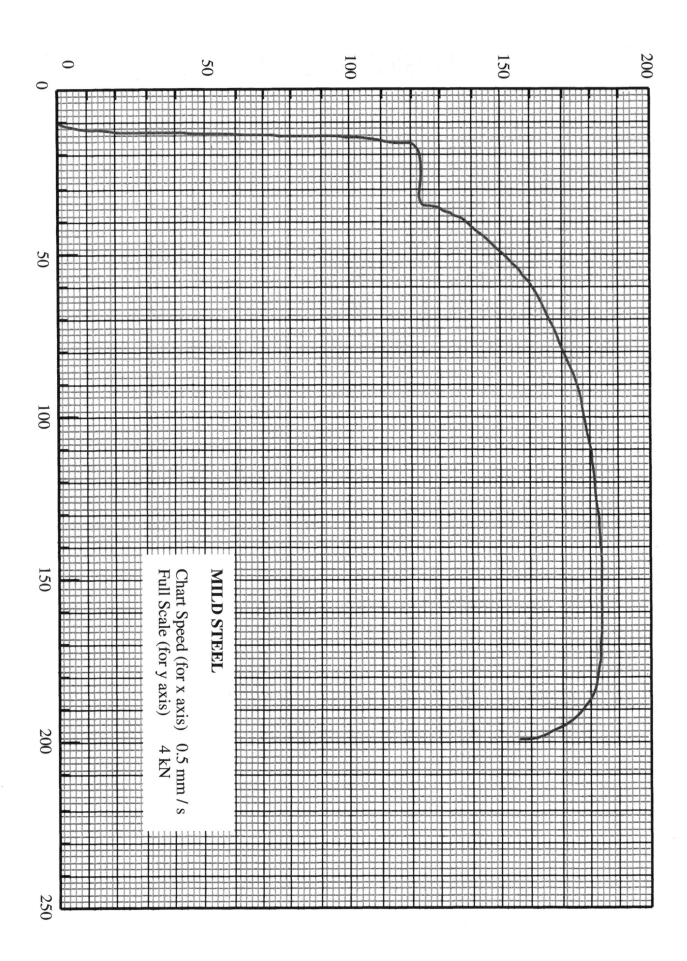

MILD STEEL

Chart Speed (for x axis)   0.5 mm / s

Full Scale (for y axis)   4 kN

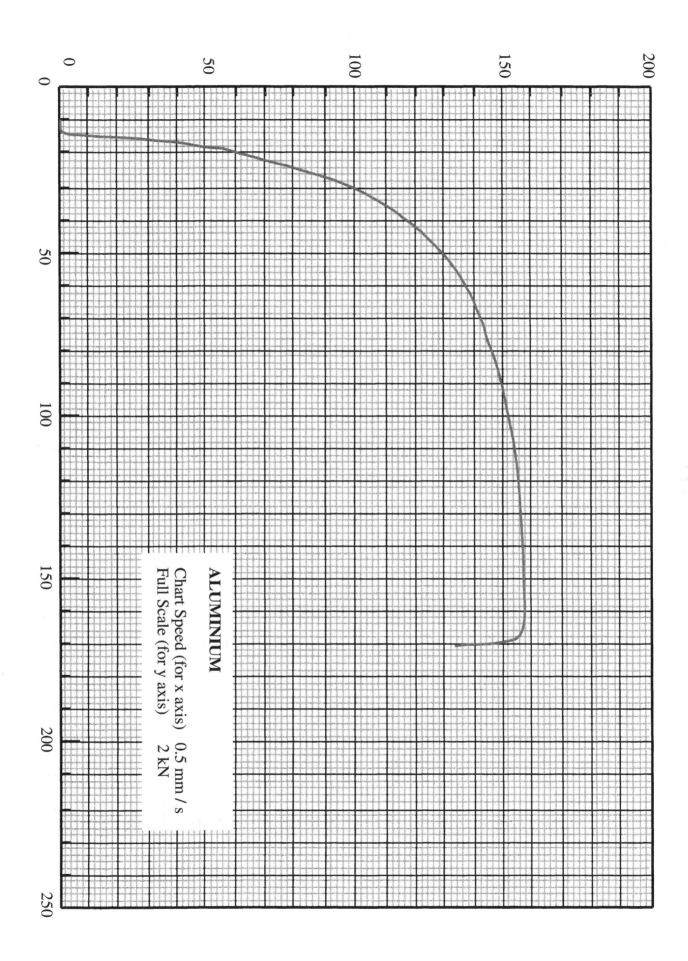

ALUMINIUM

Chart Speed (for x axis)   0.5 mm / s
Full Scale (for y axis)   2 kN

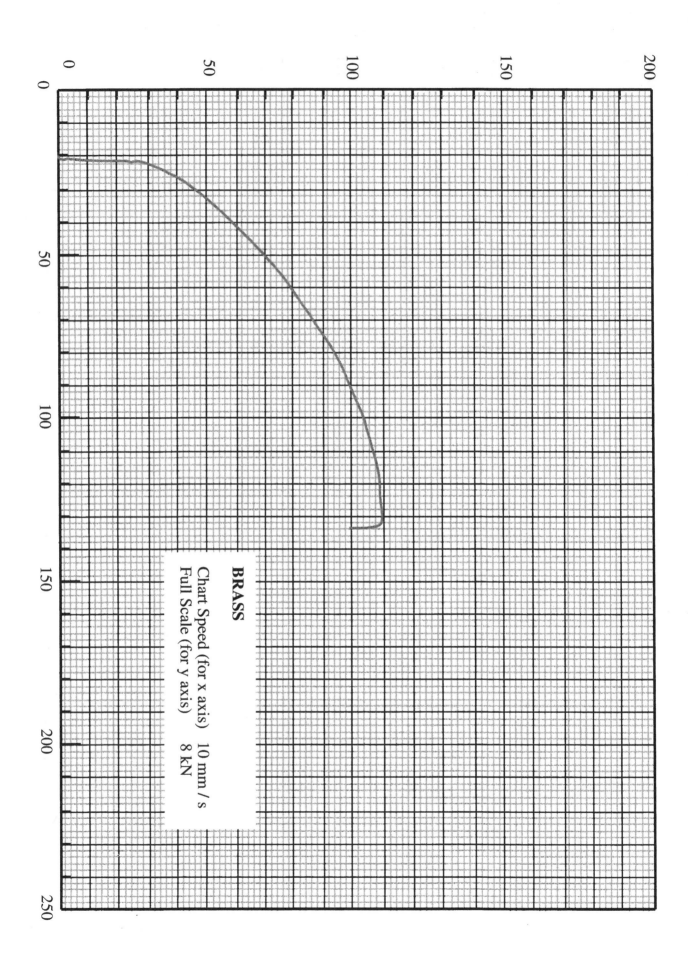

BRASS

Chart Speed (for x axis)   10 mm / s
Full Scale (for y axis)   8 kN

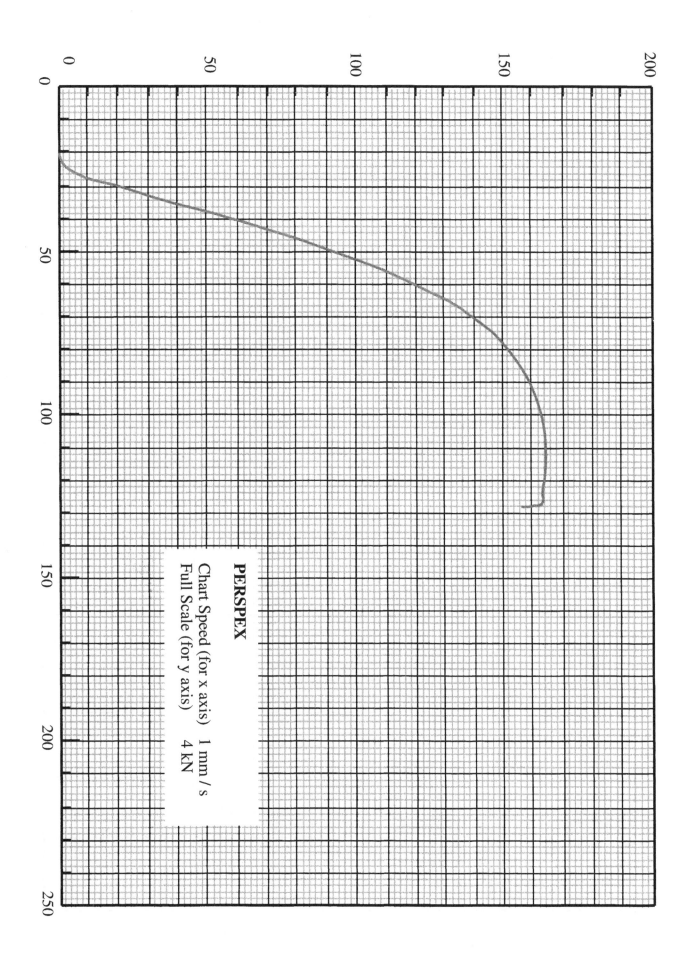

PERSPEX

Chart Speed (for x axis)   1 mm / s

Full Scale (for y axis)   4 kN

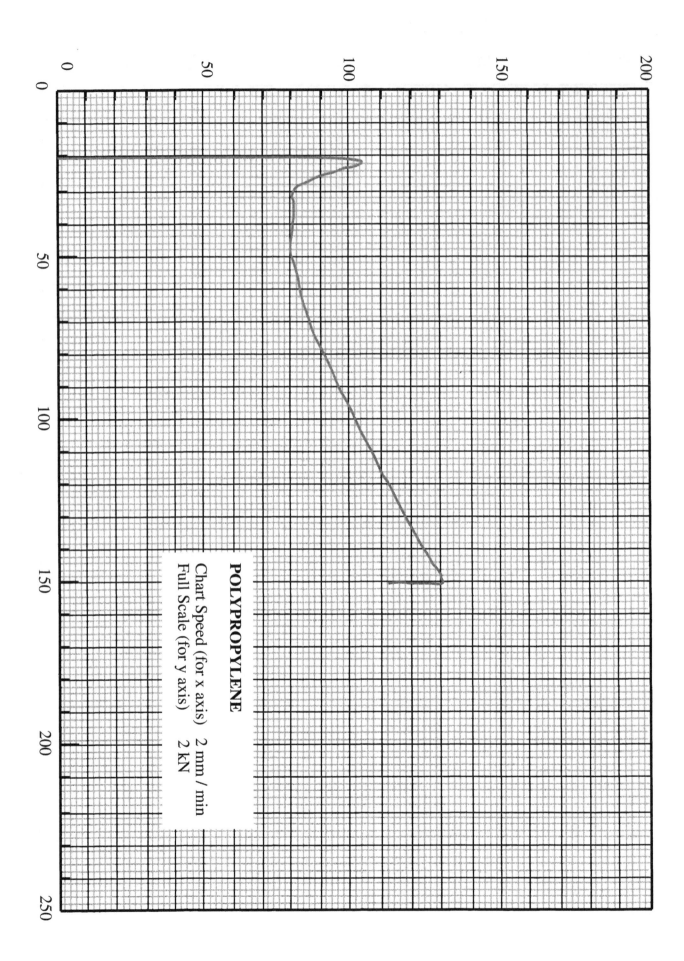

POLYPROPYLENE

Chart Speed (for x axis)   2 mm / min
Full Scale (for y axis)   2 kN

GLOSSARY AND DATA SHEET

The glossary makes an attempt to explain some of the more technical terms used in the school pack. It is not intended to give a thorough or rigorous definition of any word or term.

Amorphous	Disordered form of solid, i.e. glass-like state with no crystalline structure. Examples glass, rubber, some thermosets *(See section 1.2)*
Annealing	Heat treatment used to bring the microstructure closer to equilibrium. For example heating copper to ≈ 400 °C to remove effects of work hardening (i.e. removal of dislocations) and therefore reduce hardness and strength. *(See section 1.4)*
Atomic force microscopy/ Scanning tunnelling microscopy	Images atomic positions by measuring the force exerted on a fine needle traversed across a sample. *(See section 1.3.5)*
b.c.c.	Body-centred cubic: a crystal structure commonly adopted by metals. *(See section 1.2.1)*
Birefringence	The difference between the largest and smallest refractive index in a material, e.g. in a glassy polymer with molecular orientation, or in a stretched semi-crystalline polymer. A birefringent material has a non-zero value of birefringence. *(See section 2.8)*
Brittle failure	Failure mode which involves very little/no plastic deformation. Typical features include an overall flat/shiny surface made up of (very) small smooth areas (cleavage facets). *(See section 2.2/2.7)*
Casting	Process of pouring a molten material into a mould and allowing it to solidify. *(See section 3.3)*
Ceramic	A crystalline solid produced by the action of heat on a single or a mixture of crystalline inorganic non-metallic materials. *(See section 1.1.3)*
Cleavage	Brittle fracture along particular crystallographic planes in the grains of a material. *(See section 2.7)*
Composite	A material which is made of two or more different types of component materials in an intimate mixture. *(See section 1.1.4)*
Covalent bonding	Type of bonding in which sharing electrons gives participating atoms perceived full quantum shells. Pairs of electrons go into low energy bonding orbital and hold atoms together. *(See section 1.2.)*
Creep	Resistance of a material to deformation and failure when subjected to a static load (stress) below the yield strength at a high temperature, relative to its melting temperature. *(See section 2.6)*
Creep rate	The rate at which a material continues to elongate as a function of time when a stress is applied at a high temperature, relative to its melting temperature. *(See section 2.6)*
Crystal	A solid in which the atoms are arranged in a regular and repetitive manner in 3 dimensions. *(See section 1.2)*
Dendrite	Structure that can form as a liquid solidifies - describes the shape of the solidifying phase, i.e. 'tree-like', generally seen in metals. *(See section 1.3.1)*

Dislocation	A line discontinuity in a crystal which propagates under the influence of an applied stress to produce plastic deformation. May be considered as an extra half plane of atoms in a structure. *(See section 1.4)*
Drawing	Drawing of an amorphous/semi-crystalline polymer is where permanent deformation, due to the application of a stress, results in the alignment of polymer chains increasing the overall strength of the polymer. *(See section 1.2.2)*
Ductile fracture	Failure mode which involves a significant amount of plastic deformation. Typical features include void formation around inclusions/particles and a rough/dull surface. *(See section 2.2/2.7)*
Elastic deformation	Deformation of the material which is recovered when the applied load is removed. *(See section 2.1/2.2)*
Electron diffraction pattern	Image obtained during transmission electron microscopy and used to reveal information regarding crystal structure. Electrons are diffracted by the crystal structure in a similar way to light being diffracted by a diffraction grating. *(See section 1.2.1)*
Equiaxed	An equiaxed grain structure is one where the grains are approximately spherical. *(See section 1.3.1)*
Extrusion	Process of producing rods, tubes and various solid or hollow sections, by forcing suitable material through a die by means of a ram. *(See section 3.1)*
Fatigue	Process of progressive slow crack growth, and eventual failure, under cyclic loading - it can occur even in an initially unflawed material. *(See section 2.5)*
f.c.c.	Face-centred cubic: a close packed crystal structure commonly adopted by metals. *(See section 1.2.1)*
Flexural rigidity	The quantity EI which describes the resistance of a structure to bending, i.e. takes into account both material stiffness and beam shape. Where E is Young's Modulus and I is the second moment of area. *(See section 2.1)*
Fluidity	Measure of a liquid materials ability to fill a mould. *(See section 3.3)*
Fractography	The study of fracture surfaces. *(See section 2.7)*
Fracture Strength, σ_F	Force (N) divided by original cross sectional area (mm^2) of the sample at the point of fracture during a tensile test. Measured in MPa where $1 \text{ MPa} = 1 \text{ N/mm}^2$ *(See section 2.2)*
Glass	A glass has an amorphous structure and is often based on silica, it is not a true ceramic although shares many similar properties. *(See section 1.1.3)*
Glass transition temperature	The temperature at which a glassy polymer becomes a melt, or at which a thermoset becomes a rubber. *(See section 2.4)*
Grain	A region where the crystal structure is continuous, i.e. a single crystal. *(See section 1.3.1)*
Grain boundary	A planar defect representing the boundary between two grains. The lattice has a different orientation on either side of the grain boundary. *(See section 1.3.1)*
Hardness	A measure of a materials resistance to surface indentation or/and scratching. *(See section 2.3)*
h.c.p. or c.p.h.	Close packed hexagonal: a crystal structure commonly adopted by metals. *(See section 1.2.1)*
High resolution electron microscopy (HREM)	A form of transmission electron microscopy that allows the imaging of the atomic structure. *(See section 1.3.5)*

Hooke's Law isotropic	Stress (σ) is proportional to strain (e) during elastic deformation of some materials. *(See section 2.1)* $$\sigma = Ee$$ where E is the modulus of elasticity, or Young's modulus.
Ionic bonding	Type of bonding in which a permanent gift of electrons makes two atoms into ions with inert gas electronic structures whose opposite charges hold them together. *(See section 1.2)*
Injection moulding	Processing route used for polymers where the hot molten polymer is forced into a closed mould under pressure where it then solidifies. *(See section 1.3.2/3.2)*
Lamella	'Plate-like' shape that a phase may adopt. *(See section 1.3.1/2)*
Metal	A solid or liquid held together by metallic bonding. Shows characteristic properties of high reflectivity, high electrical and thermal conductivity and relatively high density compared with non-metals. *(See section 1.1.1)*
Metallic bonding	Bonding caused by shared bonding electrons free to flow through solid. Caused by and causes a high packing density of atoms, which is non-directional. *(See section 1.2.1)*
Microstructure	The arrangement of phases making up a solid material, generally considered on the scale of $1 \rightarrow 1000$ μm. *(See section 1.3)*
Nucleation site	Point at which a phase/precipitate starts to grow, generally a favourable low energy site, e.g. grain boundary, inclusion, vacancy. *(See section 1.3.2)*
Phase	A physically distinct form of a given material (i.e. having distinct chemical composition or/and physical structure). A material can exist as different phases under different conditions of temperature or pressure. The phases may or may not be related in some way. *(See section 1.3.1/3.5)*
Phase transformation	Denotes a change in phase caused by a change in temperature or/and pressure. *(See section 1.3.1/3.5)*
Plastic deformation	Permanent deformation of the material caused by the permanent rearrangement of the relative atomic positions. *(See section 3.2)*
Polarised light	Light filtered by a Polariser, so that the direction of the transverse electric vector of the waves lies in a single plane *(See section 1.2.2)*
Polycrystalline	Material comprised of more than one grain. *(See section 1.3.1)*
Polymer	A solid or liquid consisting of large molecules usually consisting of a carbon chain. *(See section 1.1.2)*
Precipitate	Small particle that forms in a material due to a heat treatment process. Often increases strength of the material. *(See section 1.3.1)*
Quench	A rapid cooling of a sample - may be intended to avoid a phase transformation or create a non-equilibrium structure. *(See section 3.5)*
Refractive index	The ratio of the velocity of light in a material to the velocity of light in a vacuum. A dimensionless number.
Rubber	A lightly cross-linked polymer, above its glass transition temperature, so the network chains continually change shape. Low modulus solid capable of high elastic strains. *(See section 1.1.2)*
Scanning electron microscopy (SEM)	A beam of electrons is scanned over the surface of a material. The resultant interactions can be used to form different types of images of the material surface. *(See section 1.3.5)*
Semi-crystalline	A polymer which consists of a mixture of crystals and an amorphous phase. The % crystallinity specifies the volume % of crystals *(See section 1.2.2)*

Sintering	A process route for producing ceramics where small particles/ powder is heated under pressure to form a larger solid mass. *(See section 1.3.3)*
Solute	An atom of a different species from that comprising the lattice. *(See section 1.4)*
Spherulite	'Little sphere' - the near-spherical growth form of crystals and amorphous material in most semi-crystalline polymers. *(See section 1.3.2)*
Stiffness	The stiffness of a material is its ability to resist elastic deformation and is measured in terms of its Youngs modulus, E, obtained from a tensile test and is a material property (i.e. constant for a material).Note that the stiffness of a structure is dependent on the geometry of the structure. *(See section 2.1)*
Strain	Defined by $e = (l - l_0)/l_0$. Where l_0 is the original length of the sample and l is the final length. *(See section 2.2)*
Strain-rate	The rate of change of strain with time. A material may behave very differently if it is slowly pressed into shape rather than deformed rapidly into a shape by an impact blow. *(See section 2.2/4)*
Stress	Load (N) divided by area (mm^2), measured in MPa, where 1 MPa = 1 N/mm^2 *(See section 2.2)*
Striations	Markings generally seen on a fatigue fracture surface. One striation typically formed each loading/unloading cycle. *(See section 2.7)*
Tensile strength, σ_{TS}	Maximum force (N) experienced by a sample during a tensile test divided by original cross sectional area (mm^2). Measured in MPa where 1 MPa = 1 N/mm^2 *(See section 2.2)*
Thermoplastic	A polymer which melts on heating and therefore can be processed into shape. The molecules consist of individual chains of C atoms (with sometimes O or N in the chain). *(See section 1.1.2)*
Thermoset	A cross-linked polymer created in its final shape. When used below its glass transition temperature the network chains are fixed in shape. *(See section 1.1.2)*
Toughness	A measure of a materials resistance to fracture and can be measured by the energy absorbed during a fracture test such as a Charpy impact test. *(See section 2.4)*
Transmission electron microscopy (TEM)	Transmission of electrons through thin sections of material to reveal microstructure. *(See section 1.3.5)*
Vacancy	A site unoccupied by an atom or ion in a crystal lattice. *(See section 1.4)*
Van der Waals bond	A weak chemical bond depending on oscillating mutually induced electron dipoles. *(See section 1.2.2)*
Witness marks	Marks found on a component that has been injection moulded. Caused by the join in the two halves of the mould. *(See section 3.2)*
Work Hardening	Occurs after yielding when dislocations interact to form obstacles which inhibit further dislocation motion thereby effectively strengthening the alloy *(See section 1.4)*
Yield Strength, σ_y	Force (N) divided by original cross sectional area (mm^2) of the sample when plastic deformation starts to occur during a tensile test. Measured in MPa where 1 MPa = 1 N/mm^2 *(See section 2.2)*
Young's modulus, E	Materials property: constant of proportionality between engineering stress, σ, and strain, e, in Hooke's Law. For a material which obeys Hooke's Law; *(See section 2.1)*

$$\sigma = Ee. \quad (E \text{ has same units as stress}).$$

Commonly used data and conversion information:

$0 \,°C = 273 \, K$

$1 \, MPa = 1 \, N \, mm^{-2}$

$1 \, N = 1 \, kg \, m \, s^{-2}$

	Yield strength, MPa	Tensile strength, MPa	Young's Modulus, GPa	Density Mg m^{-3}	Melting point, K
Metals:					
Aluminium	20	100	70	2.7	933
Aluminium alloys	100-600	300-700	70-80	2.6-2.9	≈830
Copper	60	400	120	8.9	1356
Brass (Cu-30Zn)	450	550	100	8.5	≈1300
Iron	50	200	200	7.9	1809
Steels	220-2000	400-2500	200	7.8	≈1750
Polymers:					
Nylons	50-90	100	2-4	1.1	340-380*
Polycarbonate	55	60	2-6	1.2	400*
Polyethylene	6-30	20-40	0.2-0.7	0.9	300*
Engineering ceramics:					
Al$_2$O$_3$	5000	-	390	3.9	2323
Carbon fibre	-	2200	390	1.9	3800
Diamond	50000	-	1000	3.5	4000
Glass fibre	-	1400-3500	76	2.6	-
SiC	1000	-	450	3.2	3110
Composites:					
GFRP 60% unidirectional‡	-	550	30	1.6	
GFRP 80% unidirectional‡	-	1200	50	2.0	
Other materials:					
Concrete	20-30	(compression)	50	2.5	900*
Wood - (Spruce)	45	(compression)	13	0.5	450*
(Balsa)	15.5	(compression)	3.2	0.18	
(Beech)	52	(compression)	10.1	0.69	

* Softening or degradation temperature

‡ GFRP 60% unidirectional - Graphite fibre reinforced plastic with 60% fibres by volume with the fibres aligned in one direction only (mechanical measurements taken parallel to the fibres).

Price per unit weight of materials and products:

(reproduced by kind permission of Professor M. Ashby)

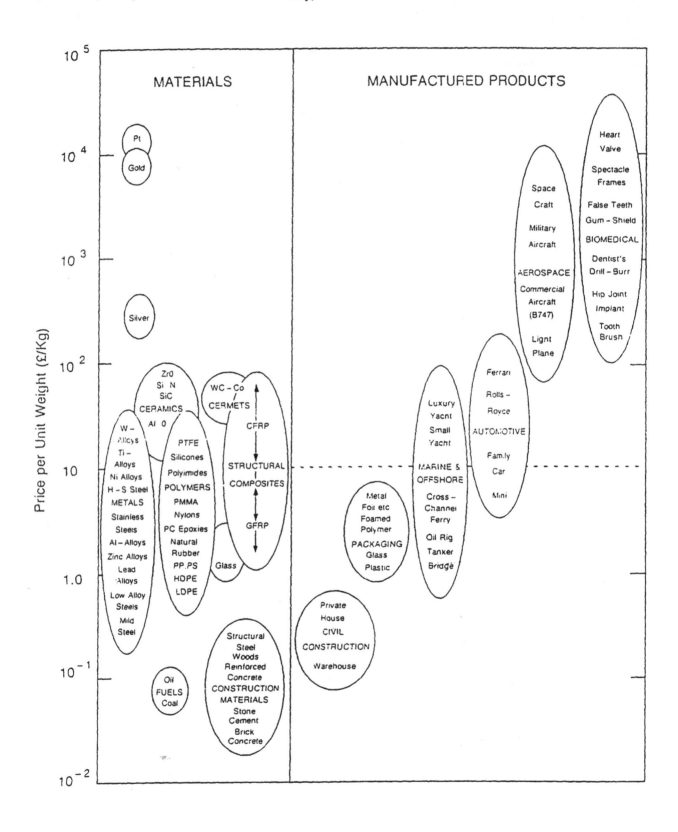

GENERAL RESOURCES

The general resources that can be used to supplement this pack are the Departments of Metallurgy/Materials Science across the country. Listed below are the Universities which offer an accredited degree course and the contact names of the people involved in the admissions area. They should either be able to help you or pass you on to someone who can. In the event of any difficulties please don't hesitate to contact me, Claire Davis, at the University of Birmingham.

Other sources of information that may be of particular relevance are listed below. Obviously I cannot include all the resources available, more comprehensive general resource lists can be obtained from the Chemical Industry Education Centre or The Institute of Materials at the addresses given below.

Bath University	Dr Tim Mays	01225 826588
Birmingham University	Dr Claire Davis	0121 414 5174
Brunel University	Dr Lynn Gabrielson	01895 203253
Cambridge University	Dr Kevin Knowles	01223 334312
Imperial College	Dr Henry McShane	0171 5946754
Leeds University	Dr Bob Cochrane	0113 2332359
Liverpool University	Dr Peter Fox	0151 7944670
Loughborough University	Dr Richard Heath	01509 223330
UMIST/Manchester University	Dr Colin Leach	0161 2003740
Newcastle University	Dr Jim White	0191 2227906
North London University	Mr Mike O'Brien	0171 7535128
Nottingham University	Dr Brian Noble	0115 9513745
Open University*	Dr Mark Endean	01908 653386
Oxford University	Dr Chris Grosvenor	01865 273761
Plymouth University	Mr David Short	01752 232637
Queen Mary & Westfield College	Dr James Busfield	0171 9755150
Sheffield University	Dr John Parker	0114 2765514
Sheffield Hallam University	Dr Malcolm Denman	0114 2533386
Strathclyde University	Prof. Alan Hendry	0141 5534152
Surrey University	Dr Mark Whiting	01483 2599611
University of Wales at Swansea	Dr Chris Arnold	01792 295749

* Candidates with Open University degrees will be assessed for professional membership and Chartered Engineer status on an individual basis.

The Education Department
The Institute of Materials
1 Carlton House Terrace
London
SW1Y 5DB

Tel. 0171 451 7326
Fax. 0171 823 1638

The Institute of Materials has details of publications, leaflets, videos and teachers courses including:

Finding out about ceramics, plastics, metals
Booklet detailing teaching resources. (1996 Edition)

Careers – Information on career opportunities in Materials Science and Engineering.

Polymer Study Tours – Short, free residential courses for teachers on aspects of polymer technology.

Chemical Industry Education Centre
University of York
Heslington
York
YO1 5DD

Tel. 01904 432523
Fax. 01904 434078
e-mail ciec@york.ac.uk

Dental Dilemmas – properties & testing of materials.
(published 1996)
Module of work aimed at GNVQ students but could provide information for a good case study linked to materials properties and testing. (£12)

Understanding Plastics   (published 1997)
Provides information on the chemistry and technology of plastics with useful information on the use of plastics in sport, the automotive industry, packaging, construction and on recycling. (£10)

Polymers – Physical testing
Twelve short chapters containing information on industry standard testing and a series of practical test activities for students to investigate the physical properties of polymeric materials. (£8)

Product Design – case studies in applications of plastics and rubbers
Three technology/science units which include background information, teachers notes and activities: polyurethane in sports shoes, sleeping bags, plastic pop bottles.   (£7.50 each)

Partners in Science Education
A directory of organisations and industrial sites that lend support.

Don Bootle
The Engineering Council
10 Maltravers Street
London
WC2R 3ER

Tel. 0171 240 7891
Fax. 0171 379 5586

Neighbourhood Engineers
The scheme attaches teams of engineers and technicians to work informally with teachers across the curriculum in their local secondary schools. A range of activities is undertaken i.e. help with project work, visits to companies, acquisition of resources, careers advice.

CREST Awards
CREST National Centre
Surrey Technology Centre
University of Surrey Research Park
Guildford
Surrey
GU2 5YH

Tel. 01483 451482
Fax. 01483 451483

Established National Award Scheme for science, technology and engineering supported by major UK businesses.

British Steel Education Service
PO Box 10
Wetherby
LS23 7EL

Tel. 01937 844443

British Steel produce numerous resources for schools. Full details can be obtained from their Education Service. One particularly useful resource would be:

British Steel GSCE Science Pack
Contains teachers' guide, pupils resource book, photocopiable student activities, samples of materials: ores used in the processes, metal strips (aluminium, brass, copper, mild steel, stainless steel). (£25)

I.G. Jones and W. Mitchell
MATTER (Project for educational software)
Department of Materials Science & Engineering
University of Liverpool
LIVERPOOL
L69 3BX

Fax. (44) (0)151 794 4675
Tel. (44) (0)151 794 5006

Materials matter in schools (published 1998)
Materials Science CD-ROM for Key Stages 3 and 4 (Needs Windows 3.1 or above)

Interactive module – it can be used by individual pupils, in small group teaching, or as a classroom demonstration. The module is in two parts. The first is entitled *What is the matter?* and addresses the following topics:
• The difference between solids, liquids and gases
• The concept of density
• Changes of state
• The difference between temperature and heat
• The meaning of absolute zero, and
• Why a solid expands on heating
The second part, *The heart of the matter*, introduces simple ideas of the atom and atomic bonding.

Printed and bound by Antony Rowe Ltd, Eastbourne

T - #0716 - 101024 - C0 - 297/210/4 - PB - 9781861250599 - Gloss Lamination